Deans of Men and the Shaping of Modern College Culture

HIGHER EDUCATION & SOCIETY

Series Editors:

Roger L. Geiger, Distinguished Professor of Education, Pennsylvania State University

Katherine Reynolds Chaddock, Professor of Higher Education Administration, University of South Carolina

This series explores the diverse intellectual dimensions, social themes, cultural contexts, and pressing political issues related to higher education. From the history of higher ed. to heated contemporary debates, topics in this field range from issues in equity, matriculation, class representation, and current educational Federal Acts, to concerns with gender and pedagogy, new media and technology, and the challenges of globalization. In this way, the series aims to highlight theories, historical developments, and contemporary endeavors that prompt critical thought and reflective action in how higher education is conceptualized and practiced in and beyond the United States.

Liberal Education for a Land of Colleges: Yale's "Reports" of 1828
By David B. Potts

Deans of Men and the Shaping of Modern College Culture
By Robert Schwartz

Deans of Men and the Shaping of Modern College Culture

Robert Schwartz

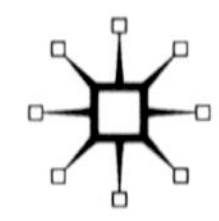

DEANS OF MEN AND THE SHAPING OF MODERN COLLEGE CULTURE

First published in 2010 by
PALGRAVE MACMILLAN®
in the United States—a division of St. Martin's Press LLC,
175 Fifth Avenue, New York, NY 10010.

Where this book is distributed in the UK, Europe and the rest of the world, this is by Palgrave Macmillan, a division of Macmillan Publishers Limited, registered in England, company number 785998, of Houndmills, Basingstoke, Hampshire RG21 6XS.

Palgrave Macmillan is the global academic imprint of the above companies and has companies and representatives throughout the world.

ISBN: 978–0–230–62258–6

Library of Congress Cataloging-in-Publication Data

Schwartz, Robert, 1950 Dec. 30–
Deans of men and the shaping of modern college culture / Robert Schwartz.
p. cm.—(Higher education & society)
ISBN 978–0–230–62258–6
1. Deans (Education) 2. College department heads. 3. Universities and colleges—Administration. I. Title.

LB2341.S316 2010
378.1′11—dc22 2010020953

A catalogue record of the book is available from the British Library.

Design by Newgen Imaging Systems (P) Ltd., Chennai, India.

First edition: November 2010

10 9 8 7 6 5 4 3 2 1

Printed in the United States of America.

Contents

Images

Cover: Reprinted from secretarial notes: Nineteenth Annual Conference of the National Association of Deans and Advisors of Men, with permission of the National Association of Student Personnel Administrators, Inc., www.NASPA.org.

1

The Rise and Demise of Deans of Men: A Historical Perspective

In the late 1800s, in the Northeast and Mid-Atlantic states, industrialization was expanding, dependence on farming as a primary vocation was ebbing, and education was emerging as a new route to success. Not surprisingly, college enrollments began to climb upward. Eager young men and an increasing number of young women saw college as a means to opportunity and success. At the same time, greater access to a college education was encouraged by a steady increase in the number of colleges. Public colleges and universities expanded quickly after the Civil War.

Beginning in the 1860s, the opportunity to attend college in America expanded geometrically in a relatively short time. As more colleges were built, offering more students the chance to go to college, dreams of a life free from the drudgery of farming and other physical labors spread rapidly among the younger generation of the late 1800s.[1] By the end of the 19th century, a new view of the future was clear. A college degree promised a better life.

The expansion was due in large part to federal money made available via the Morrill Act of 1862, which allowed states to sell federal land and use the money to construct state colleges. By 1890, a second Morrill Act expanded the law to include additional agricultural and mechanical institutions for the education of the newly freed slaves in the post Civil War era.[2]

Justin Morrill had originally intended his bill to expand the science of agriculture. Initially, agriculture was a primary feature of these public universities. But many universities quickly expanded their offerings far beyond agriculture. In 1870, President Charles Eliot of Harvard introduced the elective system, and colleges and universities soon offered students choices in class offerings that made a higher education more attractive than it had ever been before.[3]

The genius of Eliot's move to the elective system was not in giving greater choice to enrolled students but in the marketing of a college education for the recruitment of prospective students in the future. College had thus far been viewed as years of drudgery—recitation, memorization, and close supervision by faculty in often rural, cloistered, single-sex institutions. Several years spent learning Latin, Greek, Hebrew, and other "dead" languages appealed to few young men of the time. But as the promise of the elective system grew beyond Harvard, the allure of a new kind of higher education provided a powerful incentive for new students.

Beyond the public institutions created by the Morrill Acts and the freedom of choosing one's courses, huge infusions of capital from wealthy industrialists such as Carnegie, Vanderbilt, Stanford, and Rockefeller expanded the number of private colleges at the same time. In pre-colonial America, private colleges had first been founded by a disparate collection of religious groups to provide an unending supply of denominational ministers to advance each of the various churches and to expand religious training in the New World. However, after the Revolutionary War, the federal Supreme Court had ruled for Dartmouth College in 1816 and established the right of "eleemosynary" institutions such as private colleges to exist without the threat of takeover by the states. Both an immediate and long-term consequence of the Court's decision was the proliferation of private colleges.[4] Both public and private higher education spread across the country in concert with the westward migration throughout the 19th century.

As enrollments increased, so did the demand for more administration, especially for administrators who could oversee the rapidly growing student presence on campus. Prior to the Civil War, the small, "old time" colleges, as Rudolph called them, had found student supervision a relatively straightforward matter.[5] In small, rural settings, moral strictures and the drudgery of college life—reading and reciting Latin and Greek, attending to the *quadrivium* and *trivium* of the curriculum—limited the attraction of college. College presidents and a handful of faculty were able to manage the small groups of young men. Colleges were seen as extensions of boarding schools, with meals prepared under the supervision of the president's wife, minimal free time, and few extracurricular activities.[6]

But in the late 1800s, college presidents and faculty found themselves overwhelmed by students and underwhelmed by piety. As the religious sponsorship and moral inhibitions of the mid-1800s slipped away in the reformed college cultures of the post-bellum period, young men and women (the new gender on campus) were less and less inclined to bow to the authority of their elders. The combination of diminished religious influence, increased enrollments, and the rise of academic and social freedoms made for restless students.

At Harvard, the ambitious Charles Eliot found himself faced with far too many campus-based responsibilities. Seeking relief from the demands of the day-to-day monotony of university administration and eager to expand his national opportunities, Eliot appointed two Harvard faculty members to be deans in 1890. One dean, he announced, would attend to faculty concerns and issues. The other, LeBaron Russell Briggs, would be a dean "for students."[7] As professor of rhetoric, Briggs likely had taught almost all Harvard men during their first year on campus as every undergraduate was required to take Freshman Composition under his direction. Briggs' natural ease with students and his compassionate nature made him a favorite of generations of Harvard men.

As dean for students over a 40-year period, Briggs became a Harvard legend. Dean Briggs offered a compassionate ear and sound advice, qualities young men away from home and in a competitive, academic environment desperately needed. His wise counsel was eagerly sought out. When he finally retired in 1930, Briggs and his wife were surprised with a pension fund fully paid by hundreds of grateful Harvard men in appreciation for Briggs' warm friendship and sound advice over the years.[8]

Deans of men, such as Briggs, represented a new era in American higher education. Whether at Harvard or the prairie campus of the University of Illinois, many young men found themselves ill-prepared to face the academic rigors and social freedom of campus life. The first deans of men—Briggs at Harvard, Thomas Clark at Illinois, Scott Goodnight at Wisconsin, Robert Rienow at Iowa, and F. F. Bradshaw at North Carolina—were often cast into a paternal role, and acted *in loco parentis*, in the absence of parents. As such, a dean's behavior was often a function of his personality rather than originating from a scripted code of professional conduct and training.[9]

Many of the first deans were proponents of the morality of a "muscular Christianity" that emerged in the elite Eastern boarding schools of the 19th century, a pedagogy that extolled a rough and tumble approach to the education of boys and young men.[10] At the heart of muscular Christianity was the belief that rigorous play and strict discipline built moral character. By exposing boys to the challenges of the world in the safe environment of the boarding school or the "old time" college, young men learned to be responsible, honorable, and, through team play, conscious of the needs of others.[11]

In this frame of thinking, the early deans saw their role as both disciplinarian and absentee father to the young men on their campuses. Deans of men could be intimidating and even frightening to students, as a dean could invoke suspension or expulsion on a hapless student. At the same time, many of the deans of men were inclined to give second and even third chances to young men who were caught in a transgression on campus. Relying on their intuition and experience to recognize poor judgment as separate from a pathologically inclined criminal, most deans of men were stern but not mean.

Nonetheless, the position of dean of men followed a slow trajectory of development. From Briggs' transitional appointment in 1890 through the end of World War I, the few deans of men who were appointed had to follow their own best guess as to what the position of dean was intended to be. No professional associations or common training for deans of men existed. Thomas Clark of Illinois, the first dean of men to carry the actual title, described his own philosophy as a dean of men to be that of a "disciplinary officer" who was "[to] keep closely in touch and sympathy with student life and student activities. [The dean] must be willing to praise the virtuous, to commend the worthy as well as to pass judgment on the derelict."[12]

Deans were often left to make their decisions about student conduct and behavior with little direction from either presidents or trustees of their colleges. Even when they asked, few deans found definitive answers. Stanley Coulter, the first dean of men at Purdue, lamented that the deans traveled in uncharted territory.

> What is a Dean of Men?...When the Board of Trustees elected me Dean of Men, I wrote to them very respectfully and asked them to give me the duties of the Dean of Men. They wrote back that they did not know what they were but when I found out to let them know. I worked all the rest of the year trying to find out. I discovered that every unpleasant task that the president or the faculty did not want to do was my task. I was convinced that the Dean of Men's office was intended as the dumping ground of all unpleasant things.[13]

For many deans of men, discipline quickly became the most visible part of their job. Discipline of male students in college from the 1890s to the 1930s covered a myriad of sins from cutting classes and cheating on exams to gambling, drinking, consorting with women, playing cards, and riding in automobiles. A dean's office could be a busy place as Clark described in his book, *Discipline and the Derelict* (1928). Successful deans often relied on intuitive response rather than a strict interpretation of the rules and regulations. Robert Rienow, dean of men at the University of Iowa, noted, "For thirteen years I kidded myself that I was not a disciplinarian. But in the last few years I have learned that I was [always] pulling the chestnuts out of the fire for someone else."[14]

A NATIONAL ASSOCIATION FOR DEANS OF MEN

Scott Goodnight, dean of men at the University of Wisconsin, organized the first formal gathering of the deans of men in 1919. Frustrated by a lack of regular communication with other deans, Goodnight acted on a suggestion from Robert Rienow, dean at the University of Iowa, to call a meeting.

> So without authorization from anybody, I wired Minnesota, Iowa, Illinois, Indiana and Michigan to come on over. The idea of founding a permanent organization or creating a professional association was the farthest thing from my mind when I invited the boys to come in for a weekend so that we might discuss our common tribulations more intimately. It was after the first meeting had proved so pleasant and stimulating that the proposal was made to repeat it.[15]

In a letter to Thomas Clark at Illinois dated December 23, 1918, Goodnight referred to the session as "... a little conference of Deans of Men...for a discussion of our problems."[16] The meeting was planned for January 24 and 25, 1919, in Madison, Wisconsin. An agenda sent to each invited dean listed sharing their "blank forms" used by the dean to record personal information on students; a formal address by Professor L.A. Strauss, chairman of the Faculty Senate Committee on Student Affairs at the University of Michigan; and a discussion on student activities. In a follow-up letter on January 15, 1919, Goodnight assured Clark, "our sittings will be completely informal."[17]

Winter weather in Wisconsin in mid-January may have deterred some of the invited deans, as only six attended the first meeting despite Goodnight's best efforts. The deans who did attend represented the Universities of Wisconsin, Michigan, Iowa, Minnesota, Syracuse, and Iowa State Teachers College. At the time, the University of Michigan had no dean of men but on his return, Professor Strauss was so impressed that he recommended "a dean of men be appointed as soon as possible."

A second meeting was held a year later on February 20 and 21, 1920, in Urbana, Illinois, and hosted by Dean Thomas Arkle Clark. This time, deans from as far away as the Universities of Washington, and Kentucky as well as the Midwest schools, Minnesota, Iowa, Michigan, Syracuse, Indiana, Iowa State, and Grinnell College (Iowa), were invited to attend. Clark added Purdue and Pennsylvania to the list and noted that President Judson of the University of Chicago had promised he would "send one of his deans" as well. Professor Strauss at Michigan asked that his friend Dean Priest of Washington State be invited as well.

Despite the increased invitations, only eight deans attended in 1920. The group included Clark and his assistant; Deans Edmondson of Indiana; Rietz of Iowa State; Melcher of Kentucky; Strauss of Michigan; Nicholson of Minnesota; Coulter of Purdue; and Goodnight of Wisconsin. An executive committee of three was appointed to plan the time and location of future conferences.

The agenda included presentations on "The Fraternity Situation" by Dean Clark; "The Results of A Survey of a Health Supervision System" by Dean Rienow; "Student Standards" by Dean Coulter; "Student Self-Government" by Dean Edmondson; "The Housing Problem" by Dean Priest; "The Support of Student Activities" by Dean Nicholson; and "The Maintenance of Scholarship Standards" by Dean Rienow. Despite the robust agenda sent ahead of the meeting, several speakers, including Priest and Rienow, who accounted for 25% of the program, were unable to attend at the last minute. Nonetheless, the deans who were in attendance declared the second meeting a success and made plans to continue the annual event.

Scott Goodnight of Wisconsin later recalled that these first meetings of the deans of men served to reassure those in office.

> The institution of deans of men was really new in this country. There were very few of us, and we did not know of others. We were trying to educate each other as effectively as we could. [What was important was] the smallness of the circle and the intimacy with which we laid our hearts bare to each other...we had been through the great demoralization of the war [WWI] and we were trying to help effect reconstruction in the universities.[18]

THE COLLEGE FRATERNITY AND THE DEAN

Fraternity affiliation was a natural affinity group for many, if not all, of the first deans of men. In the late 1800s and early 1900s, college fraternities were an important threshold experience for young men new to the campus. Rural youth immigrating into college culture from farms and small towns learned key social skills and survival tips from their peers. College-wide orientation programs were rare, so social fraternities, such as the eating clubs of the Ivy League, were a common ground for new students. Fraternities often served as a focal point for out-of-class activities and served many young men as restaurant, boarding house, and social hub. Young men were given guidance, support, and invaluable advice on how to survive in college through the fraternity. The benefits of the fraternity were many and even extended beyond the freshman year to summer employment and career contacts after graduation.

It is not surprising then that many of the early deans of men, who were generally products of the fraternity system themselves, saw the college fraternity as an asset to their work on campus. Through a few visits to key fraternity meetings or to individual chapters, the dean of men could quickly harness the attention of hundreds of young men. Information was passed quickly through the fraternities to individual students. Even those men who were not members of the fraternities could be touched by the shaping of the campus culture at large, which was often controlled by those in the fraternal organizations. The temper of the campus was as much controlled by fraternities as it was the dean's office.[19] Standards for social propriety, even to the extent of using the fraternity men to reveal transgressions by other groups or individuals, were all at the fingertips of the dean of men, if he effectively chose to make use of the power he controlled.

So many of the deans of men who attended the first meetings of the developing National Association of Deans of Men (NADM) were members of social fraternities that it was common practice to record the fraternity affiliation of each dean in attendance. The minutes of the first 10 meetings record the fraternity for each dean. As further acknowledgement of the symbiotic relationship between the deans and the campus fraternities, a number of social gatherings ("smokers") during the first deans meetings were in located in fraternity chapter houses.

Personality as Professionalism

At the annual meetings of the National Association of Deans of Men, ruminations and anecdotal advice from pioneer deans of men such as Clark of Illinois, Goodnight of Wisconsin, Rienow of Iowa, and Coulter of Purdue, served as the training ground for younger, aspiring deans. These men, like alumni in a college fraternity, possessed an abundance of experience and knowledge. Stories from their collective past experiences comprised a major portion of the minutes of the first decade of NADM conferences from 1921 to 1931. The stories and recollections were an oral history that charted the growth and development of the new profession of dean of men. The stories also served as a road map for those men who came after the early deans. This process of story-telling also served as the primary source for the "cult of personality" that defined the role, function, and character of deans of men over the next several decades.

Between the period of the first deans of men meetings in 1919 and 1920 to the post-World War II era of the 1950s, the primary training for an aspiring dean of men was to attend the annual meetings of the National Association of Deans of Men. The other, acknowledged source of education for young deans was to be an apprentice to a senior dean, preferably one who was also a member of the NADM. Even at that, a young man who aspired to be a dean of men would need to demonstrate that he had the appropriate personality and interpersonal skills to suit the profession. As Stanley Coulter, the Dean of Men at Purdue, remarked in a speech on "The Function of the Dean of Men in the State University," "it is utterly impossible to tell what the function of the Dean of Men may be. He is a personality, not an officer."[20]

Codified through the national association, deans of men, both young and old, collectively believed that a man's personality was the key element in becoming a dean. At the time, most deans of men believed that a man's personality and his natural affinity for the work of the dean's office were far more important than any type of graduate education or training. In short, a young man might earn an undergraduate degree and then apprentice himself to a dean of men for several years in hopes of learning about and understanding the profession. In many respects, this time-honored tradition of apprenticeship was used in other professions, most notably medicine and law, until the turn of the twentieth century and the gradual rise in graduate training and education and the demand for professionalism as espoused by Flexner and others.[21]

At the time, especially in the 1920s and 1930s, this professional philosophy put the deans of men in direct opposition to the deans of women. The deans of women had, in contrast to the deans of men, embraced the idea of graduate training for current as well as aspiring deans of women. A graduate program specifically designed for deans of women was first instituted in 1916 at Teachers College, Columbia University.[22] The program was developed at the request of a group of deans of women who had attended summer classes for several years but were frustrated that they could not find the course content they felt necessary to assist them in their jobs as deans. Much of the Teacher College curriculum was developed specifically for teaching and teachers, whereas the deans of women were eager to learn more about administrative responsibilities, developmental issues among young adults, and similar issues. Soon after the development of the graduate courses for deans of women, the National Association of Deans of Women formally began.

To the deans of men, the antithesis of good practice as a dean of men was to become too formal or mechanical in one's approach to students. Stanley Coulter of Purdue lamented, "the first time I met with the Deans of Men was at Illinois [1920] and we discussed the same problems as now...We had little of mechanical devices for solving these problems. Today we have so surrounded ourselves with mechanical records that we may have ceased being personalities and have become machines."[23] The deans were against any artificial substitution or process that dehumanized the collegiate experience. In particular, the deans were opposed to the newer social sciences, such as psychology, which sought to apply personality inventories and uniform testing and measurements to groups of students. Though some deans saw the benefit of gathering knowledge and information on students, most were adamant in their opposition.

To a man, the deans of men thrived on the challenges they faced. Each day could bring a new problem to be solved or a new student in distress to the dean's door. As each problem was resolved, as each student was advised, as each crisis was averted, the dean of men's repertoire expanded. Though the solutions may not always have been neat or clean, it was the person-to-person interaction the deans reveled in and cherished. In the culture of the deans of men, it was through this process of "trial by fire" that a dean of men became stronger and more valuable, like tempered steel. A graduate education gained by sitting in a class room listening to lectures or reading books was a pale pretence. The best training was in the trenches in the company of a seasoned dean who could pass on his experience, accumulated wisdom, and insight to his young assistants. The next best learning venue was to attend the national conferences of the deans of men and share in the collective wisdom of a roomful of experienced deans of men.

The "Tardy sons of Hoyle": Discipline and the Dean

Although the early deans of men and women were far apart in their professional philosophies, they did share some similarities. One of the key parallels between the two emerging professions was a professional identity. Like the deans of women, the deans of men constantly sought validation for their role and function on campus. Although students readily accepted the role the dean of men or women as a substitute parent or at the least, an authority figure, deans of men were not so easily accepted by other administrators or faculty. Like men without a country, they were often on their own when it came to administrative roles and responsibilities.

The deans of men strongly resisted the notion that they were only disciplinary officers for errant or disruptive students. But in truth, discipline was a major public function of the dean's office. Some of the early deans, such as Robert Rienow of the University of Iowa, feared a loss of contact with students because he was perceived by so many students as the campus disciplinarian. Consequently, Rienow described how he tried to approach as many disciplinary matters as possible by avoiding punishment for bad behavior. Instead, he attempted to use reason, cleverness, and mediation to reach a conclusion acceptable to the university as well as the students.

At a meeting of the deans of men in 1931, Rienow recounted that several of the men's dormitories at Iowa had become "a rendezvous for drunks and all night gambling games." As dean, he met with the residents in the halls and warned the young men that "if we permit these [activities] we would lose the confidence of the people of the state." The university will be harmed, he warned and they, as students, would likely be expelled. Through a process of counseling, coercion, and mediation, he was able to get the students to "...agree to get rid of intoxicants, gambling and women and talk the problems over with me, it would go no further."[24] In short, Rienow promised not to expel the young men in return for better behavior on their part. Apparently, it worked.

Rienow's story captured exactly the sort of finesse that the deans of men believed graduate training or the "personnel movement" could never teach. Using his years of experience in working with young men as well as his innate intuitive skills, Rienow was able to develop what he called a level of "group consciousness and pride of ownership" on the part of the students. He convinced them that "their university" needed them to exhibit better behavior and more civic pride. Rienow achieved a mutually satisfactory result between students, the university, and the town without resorting to severe measures or creating undue publicity. Such insight could only be found in those rare individuals called to be a dean of men.

Discipline became more complicated in the 1920s. By the time young men returned to campuses after the horrors of World War I, the stage was set for an unprecedented period in the evolution of American higher education, a confluence of factors playing out on campuses that would not be replicated again until the mid-1960s. In the 1920s, a new "youth generation," a term coined by *Life* magazine, became the dominant force on the college campus.[25] College campuses were never quite the same again.

Much of the uproar in the 1920s began in the late 1800s and early 1900s when unprecedented numbers of young women arrived on campus. By the 1920s, in the lull of male enrollment during the war years, the proportion of women to men on college campuses had become almost equal. By 1925, women accounted for 47% of all undergraduate students in America.[26] Not surprisingly, the unprecedented numbers of young men and women together on campus meant new tasks and greater challenges for deans of men and women alike.

Student culture and a changing moral attitude changed many campuses dramatically. Photographs in *Life* magazine taken on the campuses of the 1920s showed young people dancing wildly to the Negro music called jazz, depicted wild parties, and suggested sexual improprieties. As *Life* magazines flew off the newsstands, the life of deans of women and men became even more difficult. Even smaller, private campuses felt the impact of the change. Dean William Alderman of Beloit College claimed that "deaning in the "20s," whether at Purdue, Wisconsin, or on his small campus in Beloit, Wisconsin, was a difficult job for even the heartiest of men. Public perceptions of college life in the 1920s, Alderman claimed, was of a

> ...heterogeomy of monstrosities—bobbed haired daughters of Satan in their early nicotines; tardy sons of Hoyle who have an aversion for the hardy sons of toil; emulous coeds who indulge in such fatuous anachronisms as breaking the endurance record for the tango;...The daughters of culture have married the sons of prosperity and are sending their offspring to universities and colleges because it is fashionable, convenient and prudential.[27]

As deans, Alderman and others knew that many, but not all, of the stories were overwrought and, to a large degree, "ginned up" to sell more magazines. Dean Blayney of Carleton College in Minnesota noted what *Life* magazine and the public did not see. "Ideality and frugality still haunt college halls, and...students continue to bring their hopes and poems to their deans and professors."[28] Blayney urged, "deans of men [must continue to]...be all things to all men."[29] In particular, Blayney argued, "... The deans, especially deans of men, are quite universally considered as being primarily interested in business and administrative matter, rather than in scholastic matters."[30] He urged that social and disciplinary issues were too great a focus of the dean's time and energy. "College deans...should form...a part of the shock-troops in the great struggle against inefficiency and superficiality in [American] higher education"[31]

When the dust settled and the smoke cleared, the deans of men successfully survived the turmoil of the 1920s. The re-emergence of the dean of men and his role on campus was further evidence to the deans of the NADM that their commitment to the "cult of personality" instead of graduate education was validated beyond dispute. There could be no greater test of the position of dean of men than the decade of the Roaring '20s. Deans of men had survived a great test.

At the 1931 deans' conference in Gatlinburg, Tennessee, Joseph Bursely, dean of men at the University of Michigan, reiterated the common theme. "I am afraid that I am not in sympathy with the idea of any course of training for the position of Dean of Men.... The best and most successful Deans of Men are born and not made."[32]

However, Bursely continued,

> there is one place where I believe preparedness is absolutely essential to the success of a dean of men—that is in the selection of a wife. The very best preparation [a dean] can have is to marry the right woman. If she is the right kind, a dean's wife does just as much to earn his salary as he does, and if she is not, he might as well quit before he starts.[33]

A FORK IN THE ROAD?

Few dissenting voices were heard at the NADM meetings. However, one voice that did counter the common beliefs of the other deans of men belonged to F. F. Bradshaw, dean of men at the University of North Carolina. Bradshaw became a convert to the new "student personnel movement" and ran counter to the "deans of men are born, not made" philosophy that guided the deans of men.

The student personnel movement began at Northwestern University in Evanston, Illinois, when psychologist Walter Dill Scott became president of the university.[34] Scott had worked for years on improving and increasing efficiency in business and industry through his theories of applied psychology. Scott believed that by developing a personality inventory of employees, good supervisors could categorize employees based on their character traits, characteristics, work habits, and personality. With these inventories in hand, jobs and work settings could be matched or at least, more closely aligned. As a result, increased efficiency and higher rates of job satisfaction would be achieved.

Scott first applied his techniques in industrial settings such as factories, where work and employees were often matched by job vacancies or demands. He later was able to realign his practice of personnel psychology, as he called it, to inductees in the military during World War I. As draftees and new recruits were assigned to various units out of basic training, Scott was able to complete psychological inventories on some and then match them by personality to open slots in the various Army units.

When Scott was named president of Northwestern, he quickly used the opportunity to apply the same principles and theories to the college student population. Scott removed the positions of dean of men and women and replaced them with new positions, which he called personnel deans, assigning one to the male students and one to the women. Students were assessed using the personality inventories developed by Scott and others, but adapted to the college population. Scott believed that colleges and universities would benefit from his techniques in much the same manner as personnel in the military and in business and industry had successfully adapted in his prior applications.

Two of Scott's protégés, L. B. Hopkins and Esther Lloyd-Jones, became nationally known through their work in the new student personnel movement at Northwestern.[35] Hopkins was in the role of dean of men and Lloyd-Jones served as his female counterpoint. Both Hopkins and Lloyd -Jones published papers and monographs and gave presentations based on the personnel experiments at Northwestern, and both gained national recognition for their work. Through their work and national reputations, both Hopkins and Lloyd-Jones advanced to other positions. Hopkins became the president of Wabash College, a single-sex, liberal arts college for men in Indiana. Lloyd-Jones joined the faculty at Teachers College, Columbia University where she joined Sarah Sturtevant and Ruth Strang in the Department of Higher Education. Lloyd-Jones' book, *Student Personnel Work at Northwestern* (1928), was one of the primary sources for the edification of many deans of women and other converts to the student personnel movement in the late 1920s and 1930s.

A New Kind of Dean of Men

F. F. Bradshaw was a fairly young man when he took over the position of dean of men at the University of North Carolina in Chapel Hill in 1931. Eager to advance himself and especially eager to gain more knowledge about his new position, Bradshaw spent two summers in New York, where he took classes at Columbia.[36] His exposure to new ideas and theories of managing students on the college campus caused him to make very interesting changes on campus in Chapel Hill, and made him a new voice at the NADM meetings.

F. F. Bradshaw was one of the first deans of men to publicly declare that there might be real value in psychological testing, interviews, record-keeping, and "objective" measures of personality as promoted in the personnel movement. Bradshaw's ideas were a rare counterpoint to the traditional discussions held at the NADM meetings. At the 1931 NADM meeting, Bradshaw was one of several speakers to address the topic, the preparation of the dean of men. Bradshaw followed Dean Bursely of Michigan, who had made his comment about marrying the right sort of woman. In his remarks, Bradshaw noted the lack of formal or graduate training available to the deans of men, and argued that "... in our deanly world we are not saved by any training processes whatsoever." This makes it a little difficult to talk about the dean of men's preparation for his work. While deans were born and not made, they "might be made better [emphasis added] by preparation."[37]

Bradshaw was particularly concerned about the quality of student advisement and counseling available on campus. Unlike his peers, Bradshaw had started to incorporate some of the new ideas he encountered at Columbia. Bradshaw argued that the use of modern psychology and other social sciences had great application for the college campus and the work of the dean of men. In closing, he made what would become a prophetic statement, "the deanship stands at a fork in the road... [the future of the dean is in becoming] administrative coordinators of... the whole individual student and of the group life of students."[38]

Despite Bradshaw's arguments, throughout the 1930s, a majority of deans of men held to the belief that the best deans of men were born, not made or, in other words, not trained through graduate education. Instead, they held, at least publicly, to the notion that a man was either born with the innate personality characteristics to be a dean like a Thomas Arkle Clark, or Scott Goodnight, or Robert Rienow, or even L.B. Briggs, or he wasn't. Fred Turner, protégée and successor to Thomas Clark at Illinois, reported on the preparation of deans of men at the same 1931 meeting where Bursely and Bradshaw both spoke. Turner had conducted a survey of deans of men on the topic and presented his findings at the conference. Reading from his survey results, Turner noted that, "[it is the] general opinion that there is no satisfactory training [to be a dean], at least from the academic standpoint, for the simple reason that the best deans are born that way and not trained that way." [39]

However, when Turner conducted a second survey seven years later, in 1936, the results had changed dramatically. Of 175 surveys mailed to deans, 128 were returned, most from men who reported that their title was dean of men.[40] A total of 104 of the deans had taught in a college or university prior to becoming a dean and another 45 were already in college administration. Asked what courses had been most valuable to prepare for their work as deans, psychology was first (62), and education second (45). Liberal arts courses (31) and sociology (30) came in third and fourth, respectively.

When Turner asked the respondents specifically, "What is your reaction to the statement that a Dean of Men is born and not made?" 25 deans in the first survey had agreed that the best deans of men were "born, not made." But by 1936, 75 deans responding to the second survey agreed that some training could be useful. Another 22 deans disagreed with the notion that deans were born to the job, and 6 did not respond to the question. A follow-up question, "what inherent qualities should a man have to be a dean?" generated an extensive list of social traits such as congeniality, good mixer, and "temperamental traits," such as "backbone," "energy," and the like as most desirable.

When asked to cite specific course work that might constitute preparation for deans of men, the suggestions included "broad liberal arts course (57);" "any academic subject as a major field (17);" and the non-committal response of "graduate work essential (10)." Education (123) and psychology (114) were the top content areas recommended. In the category of practical training, the top suggestions were "apprenticeship to a Dean of Men (68); work with [student] activities (43); administrative duties (30); counseling and interviewing (27); dormitory proctor (18); and business experience (16)." Other random suggestions included "Y.M.C.A. work, grade tests" and "speaking in public."[41]

Turner also compiled a list of "some of the courses now being taught" in the field. The list provided an interesting perspective on the status of professional preparation programs for deans of men and women in the mid-1930s. Teachers College, Columbia University, boasted 22 specific courses either directly related to the work of the dean or in education, personnel, guidance, and the like. Several other schools—Northwestern, the University of Chicago, and the

University of Wisconsin—offered summer institutes for aspiring deans. In many cases, the curricular emphasis was on vocational guidance or training.

Turner's 1936 report indicated a significant change for the deans of men. By acknowledging the array of course offerings provided by numerous institutions across the country, the deans recognized that professional training was becoming the standard, not the exception. Nonetheless, the old debate continued. Many incumbent deans of men still believed that though some professional training or graduate education might prove useful, the best deans of men were still drawn to the profession by their personality, not the addition of another degree or time in graduate school.

A Slow Change

Inevitably, the change in philosophy toward professional preparation caused a change in the direction of the deans' national association. In their meetings, the deans of men struggled to determine where to go as a profession. Harold Speight, dean of men at Swarthmore, suggested to his peers that, "...while there is no specific direction upon which we can expect general agreement in preparing men to serve as Deans and Advisers of Men, the greatest hope lies in the development of [the] profession. We should build up the profession through apprenticeships, and maintain...a list of young men...now in training."[42]

But even this seemingly innocuous plan met with opposition. Dean Massey of Tennessee countered, "I do not think that I shall ever give to the Association of Deans of Men the statement that I need an assistant. I may sit down and write to some of these men...in confidence, but I certainly wouldn't give it out where it would be broadcast—because a man who has everything on paper to make a good Dean of Men [may have] nothing at all in his clothes that makes him a good Dean of Men."[43] The notion of personality, even by 1936, remained constant. So although a young man who aspired to be an assistant to the dean or an assistant dean of men might seek professional certification through graduate study, the clear message, at least at the national meetings of the deans, was that the best avenue to success was to find a mentor. Apparently, only those young men who could find a sitting dean of men to vouch for them would be admitted to the club.

To a degree, the issues of personality versus professionalism were rooted in the beginnings of the deans of men. To violate the precepts laid down by L. B. Briggs and even more importantly, T.A. Clark and

Stanley Coulter, was the worst sort of treason. To use graduate training and education, not personality, as the core indicators of the next generation of deans of men would be a repudiation of the history of the deans. Many of the deans still in office in the mid-1930s knew Clark, Coulter, and others. Many of the men who first established the NADM, such as Scott Goodnight from Wisconsin and Robert Rienow of Iowa, were still in attendance. Too many of the deans could recall firsthand the fireside chats and smokers at fraternity houses that had characterized their early years as deans. To ignore or deny the roots of the profession and the national organization would be like disowning a family member. Personality continued to win the day over professionalism even in the late 1930s.

Much of the tone of the work of the early deans of men was not set by professional standards achieved through a common core of graduate study or a set of professional standards or an accreditation process that was defined by a national organization. Most of the deans of men who were in office at the end of the 1930s operated through a consensus of beliefs and attitudes toward their jobs, but much of the day-to-day operation depended on the incumbent in office from campus to campus. Variations abounded and variety was the rule, not the exception. Nonetheless, the deans of men had a loose but consistent credo: Teach young men to be obedient, respectful, and diligent in their studies and in their out-of-class activities.

In the religiously affiliated institutions of an earlier time, not only did piety dictate behavior, so did the intimate and closely supervised environment of the college campus. The college president and college faculty on many colleges of the 19th century would eat meals with the students, often in the same room. Student behavior was intertwined with close supervision and everyone on a college campus was in close proximity. Yet there are clear examples where even in such environments, for example, Harvard and the University of Virginia, student behavior was far from exemplary. Kathryn Moore and Jennings Waggoner have documented behavioral excesses on both campuses.[44] Food riots date back into the 1600s on early American colleges as food and the eating hall was a common event shared by all students. Poor food quality, such as the "butter wars" at Harvard, where rancid butter was served once too often, irked cantankerous students. But in general, behavior and discipline were never major concerns until the size of colleges and universities grew exponentially in the late 1800s and early twentieth centuries.

In pubic colleges and universities of the early twentieth century, students were guided by rules and regulations issued by the deans of men and women with the assent of the president and other administrators, the faculty, and by default, the public. The early deans such as Thomas Arkle Clark, Scott Goodnight, Robert Rienow, Stanley Coulter, and others, replaced the morality of religious affiliation that guided campus culture in the 19th century with a more secular, personal morality in the twentieth century. As the arbiters of student life and consequently, student behavior and discipline, the deans of men who carved out a role and place for themselves on the modern university campus found much of their approach to their work was in fact, parental. Thomas Arkle Clark, the dean of deans, was very much the paternalistic figurehead of the Illinois campus. Some students loved him, some feared him, but all students respected him in his role as dean. As such, Clark set the tone for deans of men for the first half of the twentieth century. LeBaron Russell Briggs did the same at Harvard, in many respects, but Clark, unlike Briggs, was at a modern, public, coeducational university, a model that became very familiar for many other administrators at similar institutions across the country. In many respects, Briggs was still the benevolent headmaster at an elite institution whereas Clark was an administrator at an emerging, public university. Clark's definition of the dean of men became the more relevant standard as Illinois was more relevant to other institutions and other deans. While admired, the Harvard model was relevant to only a few.

Clark, Scott Goodnight, Stanley Coulter, Robert Rienow, F. F. Bradshaw, and Fred Turner exemplify the men who served as deans of men from the first appointed dean of men through the post World War II era. But the days of the dean of men were limited. Many of the deans of men who were on campus in the post-war period of the 1940s and 1950s re-emerged in new roles or with new titles. But of the antecedents for those who served as student personnel, student development, or student services staff throughout the twentieth century and now continue to play a significant role in higher education in the twenty-first century as deans on most campuses were determined and first acted on by the deans of men.

The following chapters chronicle several of those who first served as deans of men. By examining their activities and how they approached their work, we gain a better understanding of the role and scope of deans of men from the 1900s through the 1950s. After a close examination of four deans of men who span the time between 1900 and the 1950s, a separate chapter examines the role of deans of women. A chapter is also devoted to an examination of the post World War II period of the deans of men and the juxtaposition of the roles of deans of men and women in that era. A final chapter is devoted to summarizing and re-examining the long history of deans of men and the impact they have had and continue to play in the professional life of contemporary student life and culture and on higher education in general.

2

The Pioneer: Thomas Arkle Clark, Dean of Deans

Image 1 Thomas Arkle Clark, Professor of English (1893–99) and dean of undergraduates (1901–1909) and men (1909–31). Courtesy of the University of Illinois Archives.

"A tall, spare, white-haired man with a bushy white mustache was carried into a Chicago hospital last November to undergo a serious operation. Flowers, letters, telegrams began arriving for him. 'Who is this guy?' asked an attendant. Replied another, 'Guess he's a gangster.'" This was the beginning to the 1931 *Time* magazine obituary for Thomas Arkle Clark, officially the dean of men at the University of Illinois from 1909, and by most accounts one of the first dean of men in the United States. The obituary described Clark as

> Well-beloved, well-hated, "Tommy Arkle" wore garish clothes, big rings, liked to be told that he was the best dressed man on the campus, glowered quizzically over his spectacles as he talked with his students. Quietly, firmly he made his impress upon Illinois, abolishing naughty fraternities (Kappa Beta Phi, Theta Nu Epsilon), fraternity "hell week," freshman hazing, student ownership of automobiles. He is fond of proper fraternity life, interested especially in his own Alpha Tau Omega.[1]

When Clark attended the National Association of Deans of Men conference in 1931, it was his last. Fittingly, he reflected on his long and successful career. He attributed some of his success as the dean of men at the University of Illinois to his hardscrabble childhood.

> I had to run the farm. I had to do everything.... At seventeen I had run the farm two years. Mother (actually, his aunt) was a very sensible woman; she gave me the responsibility; I made the decisions;... I think that responsibility was one of the things which prepared my soul for the job I had to fill afterwards [dean of men], as much as anything that ever came to me.[2]

BEGINNINGS

Thomas Arkle Metcalf was born in Marshall County in 1862, near Putnam in north-central Illinois. His parents, who were cousins, had immigrated from England in 1849 and married soon after they settled in America. His mother, Mary Arkle Metcalf, died when Thomas was only a few months old. His father, William, moved downstate to be closer to family members near Rantoul, Illinois. A former coal-miner, William struggled to become a successful farmer. When he died in 1872, ten-year-old Thomas became the adopted son of his aunt, Mary Metcalf Clark. He later took his adoptive parents' surname, Clark, in large part out of respect for his new family.

As a child, young Thomas Arkle's weight was a constant point of discussion. His adoptive parents often weighed him to monitor his health.[3] A scrawny child who was prone to illness, Clark was limited in his early physical activities, including farm chores, for fear that he would over-tax his frail body. His parents' early deaths no doubt exacerbated these concerns. Over time, Clark grew to be a healthy, young man but he remained thin throughout his life.

This hardscrabble beginning, according to Clark, became his early training for his work as dean of men. By his own account, "Suddenly, without any warning at all, my father died . . . I had at once the responsibility of being the head of the house, and looking after Mother, who was at that time sixty-five years of age. . . . .[4]

Clark valued his adopted parents and siblings. His personal correspondence reflected his deep appreciation for family. On his many travels, he faithfully sent postcards and letters home, describing his trip, the surrounding countryside, his demeanor and state of health, even including his weight as an inside, family joke. The cards and letters were usually signed "T.A.," his family nickname.[5] Clark stayed in close touch with his extended family, visiting frequently by car (the small towns of Turlock and Rantoul where Clark had grown up were within a 20-mile radius of the Champaign-Urbana University of Illinois campus).

A College Man

Despite his affinity for his family and home life, like many young men of his time, Clark was eager to leave the family farm in search of new career opportunities. But because of his family obligations on the farm, Clark had to delay his dream until 1886, when he persuaded his aunt/mother and brother to sell the farm and move to Champaign, where at the age of 24, he could finally realize his dream of a college education. The delay in his education made him a bit older than some of his peers when he was finally able to enroll in the University of Illinois Academy, a precursor to full enrollment in the university.[6] Despite his delay in pursuing a career, Clark's ambitions moved him quickly upward.

Studying rhetoric, Clark earned his bachelor's degree in 1890 at the age of 28. Despite his age, Clark embraced the opportunity to engage in various student activities, including membership in a fraternity, Alpha Tau Omega. He saw himself as "being a politician" but, he acknowledged, he found he enjoyed managing campaigns more than running as a candidate himself. Clark also worked for the local newspaper, the Champaign *News-Gazette.* Clark thought he might become a "newspaper man," but the first job he held after his graduation in 1890 was teaching in a local school of 500 students, Eastside, rising to become the principal. The school was located, according to Clark, in "the slum district of the town [populated by] children of the saloon keepers and the prostitutes of the town and all colors.... They all told me I would be run out in a week and I thought maybe I would."[7]

Clark taught school for three years. In 1893, he got the chance to return to the University of Illinois and quickly took the job. He began his academic career as an assistant professor of rhetoric, initially replacing a Professor Brownlee who had taken a leave of absence. Clark quickly advanced himself in the university in what would become a familiar pattern of opportunism throughout his career. After his promotion to Associate Professor in 1895, Clark spent two summers at the University of Chicago taking advanced courses to expand his academic credentials. In 1898, he attended Harvard for one year. These efforts paid off in his promotion to full professor and head of the department in 1899.[8]

PROFESSOR CLARK

In his role as a faculty member, Clark encountered many young men new to the university, often boys who had left a family farm. Most of the students at Illinois were from rural backgrounds, farm boys like Clark himself. Now in his thirties, Clark, like Briggs at Harvard, developed his intuitive skills to a high level and became a valued adviser for many of the undergraduate men at Illinois. Clark was an active member and alumni supporter of the fraternities at Illinois, especially his own Alpha Tau Omega chapter on campus. It was a natural progression for him to become involved with other student groups as well.

Clark was an ambitious man. Eager to promote himself within the university, he caught the eye of President Andrew Sloan Draper on more than one occasion. Clark's self-promotion resulted in his academic and administrative advancement first to full professor as a result of his extended course work at the University of Chicago and at Harvard. He then assumed the position of Acting Dean of the College of Literature and the Arts in 1900, a position that placed him in full view of Draper on a regular basis.

In 1901, knowing of Clark's skills in advising young men, Draper prevailed on Clark to take on the case of a difficult student. The young man, son of an influential family and a member of the baseball team, had ignored numerous efforts by Draper to attend classes on a regular basis and stay out of trouble. Frustrated, Draper told Clark, "I can do nothing with him (the student). See what you can do." Clark achieved what Draper could not—he got the young man back on track and into his classes.

Draper was sufficiently pleased (and no doubt relieved) at the young man's change in attitude that he gave Clark a new title, Dean of Undergraduates, and moved him to an office near the president's office in the library building on campus. Clark also carried the title Secretary to the President and was made a regular member of the Council of Administration, in essence, the primary administrative body of the university, a position Clark would hold until he retired in 1931.

Clark's appointment, earned primarily as a result of Draper's unrelenting support, was not without rancor. In particular, the Illinois provost, Thomas Burris, opposed several of Draper's reforms of the Illinois administration, and in particular, he objected to Clark becoming Dean for Undergraduate Students in 1901. As Burris wrote to Draper, "I will say at the outset that I cannot see the way clear to the appointment of any one man, by whatever title he may be known, who shall be expected to stand officially or otherwise between yourself and the heads of the departments"[9] Burris objected to Clark's appointment on personal grounds as well as professional. As Nidiffer and Cain note, "Burrill believed that Draper inappropriately favored Clark, provided him with unwarranted raises and promotions, and granted him authority in issues over which he should have no purview."[10]

It is clear that Clark used his good relations with Draper to every advantage he could. No doubt his efforts to curry favor with the president found objection among other faculty beyond Burris. Taking full advantage of Draper's determined reformation of the Illinois administration, even over the objections of Burrill, Clark continued his successful and ambitious rise in the ranks.

In 1904, Edward James became president of the University of Illinois, but made few administrative changes. Clark held his position until 1909, when James changed the office to Dean of Men. A parallel position of Dean of Women had been in place since 1897 according to university records, to help accommodate the small but steady increase in the enrollment of women students at Illinois. However, women were a distinct minority at Illinois.

In November 1908, Clark was approached by representatives of Stanford University, who inquired as to his interest in becoming

Dean of Men at Stanford. In February 1909 a solid offer was made. Clark communicated this information to President James, who asked Clark what it would take to keep him at Illinois. Clark's response was that it was not merely a matter of salary but of jurisdiction. In response, James appointed a special committee of the Council of Administration to consider the issues at hand. The committee recommended expanding Clark's role as Dean and no doubt, an increase in salary as well. Subsequently, Clark was named to be Dean of Men and given greater latitude in the authority of his duties. Clark's office, along with several other administrative offices, was moved to the natural history building where he took possession of a larger office, a further recognition of his expanding responsibilities.[11]

Under James's direction, Clark invested himself to eradicating several pockets of corruption, dishonesty, and repugnant behavior among the student population. One of the first challenges Clark took on was the eradication of the secret fraternity Theta Nu Epsilon, The Theta Nu Epsilon chapter at Illinois had, over time, become notoriously powerful in student politics and government. Clark wrote a scathing article in the school paper blasting the fraternity and its members for behavior that should not be tolerated on any campus. Theta Nu Epsilon threatened to sue Clark for libel but the threat was never acted upon. Instead, Clark used his influence with other students to minimize the power of Theta Nu Epsilon over time, eventually crushing the organization under the weight of public opinion.[12]

Another area that Clark and James attacked was the spoils system associated with campus publications. For years, the editors and staff of the campus yearbook and newspaper hired or appointed their friends and associates to positions of responsibility with little if any regard to talent or aptitude for the nature of the work of publishing a yearbook or writing the paper. Even more egregious were the profits made from advertising and sales, which were distributed freely to selected parties. Student publications raised as much as $1,000 per year in advertising revenues and sales, no small matter as $1,000 in 1908 would be equivalent to $240,000–250,000 in 2009.[13] Clark gradually instituted a Student Publications Board to oversee the newspaper and yearbook and to appoint editors and other staff on a merit-based system.

The third leg of reform in student life that Clark took on under President James was hazing. Like many colleges across America, hazing had become an accepted practice, perhaps carried over from boarding schools, among the largely male student populations. At the time, much of the hazing occurred between the sophomore and freshman classes, urgently incited by the upperclassmen. On many campuses, the juniors incited the freshmen and the seniors sponsored the sophomores. Often the hazing took the form of outright brawls and fist fights, although other forms of intimidation and harassment were never too far undercover.

Clark used a variety of means to reduce the hazing, just as he had other problems among students. Students caught hazing were dismissed from the university. Others, threatened by the social sanctions and dismissals, either ceased or went deep underground to avoid detection. Clark also began a studied campaign of "getting to know" the students as they came to campus. In 1909, Clark began to organize photographs of each new freshman in what some called his "rogues gallery." He also wrote to all incoming freshmen, urging them to stop by and see him and offering fatherly advice even before they arrived on campus. Clark's methods proved successful as he soon learned many of the names and faces of the students, and, over time, could call many of the students on campus by name when he passed them on the street or if they came to his office for some of his proffered advice and counsel.[14]

Like a sheriff in an old-time Western movie, Clark, in his role as Dean of Men, walked the streets and ruled by various forms of intimidation, skill, and a bit of luck. Referred to by the students as "Tommy Arkle," he used information gleaned from his student contacts to narrow the lists of miscreants and established himself as a man of infinite wisdom and wise counsel to others. He built his capital slowly but convincingly among the student population and shrewdly used the many tools at his disposal to maintain order and social control.

ALICE BROADDUS CLARK

In his personal life, Clark married Alice Broaddus on August 24, 1896.[15] She was also a graduate of the University of Illinois, was a member of Alpha Phi sorority, and served as president of both the sophomore and senior classes during her student days. Alice Broaddus earned two bachelors degrees from Illinois, one in science and one in art. Like Clark, she served as a principal of a high school for two years in Forest, Illinois. She had also enrolled in Sargent Normal School, where she pursued courses in physical training, leading to a two-year term as Director of Gymnasium at Mary Nash College in Sherman, Texas.[16]

Alice Broaddus Clark, despite her own professional talents, became best known as Mrs. Thomas Arkle Clark, the wife of an ambitious husband. She did not continue to pursue an active career of her own outside the home. Nonetheless, she was a noted public figure in Champaign and at the university for the next 30 years of her life. A childless couple, the Clarks valued their relationships with students at the University of Illinois much as they might have embraced a family of their own. Alice Clark played a key social role with students and faculty inside the university community.

These relationships were never more apparent than in the close relations developed between the Clarks and the numerous young men who served Dean Clark as assistant deans of men. Each year, Dean Clark employed several young men to attend to the numerous clerical duties in the office. As the responsibilities of the office grew, so did the number of men. Several Clark-trained assistants went on to other administrative positions at other campuses. Fred Turner, a young man who expected to go on to medical school, changed his career after his work in the dean's office. Turner eventually became Clark's successor in the mid-1930s.

When Clark passed away in 1932, Alice Clark relied on Turner not only to help with funeral arrangements, but even the decision as to the burial plot and other issues. In correspondence, Alice Clark thanked Turner for his loyal service and told him that had been the Dean's favorite.[17] Widowed when Clark died in 1932, Alice eventually moved to Peoria, Illinois, where she died at 80 on August 27, 1948.[18]

CREATING THE DEAN OF MEN

Over the course of his career at Illinois, Thomas Arkle Clark slowly created a stage for himself that extended far beyond the Illinois campus. He used the "bully pulpit" of the Dean's Office to launch an impressive campaign that culminated in a national reputation and standing. Clark established himself as not only the first dean of men at Illinois but the first dean of men nationally. Through a never-ending stream of dispositions on character and moral teachings delivered through both public appearances and the mass media of the time (radio and newspapers), Clark became a celebrity. He established a moral tone and tenor for campus life that set a pattern for the multifaceted role of deans of men, beginning first in the Midwest and then spreading nationwide.

As the first dean of men, Clark was in a unique position in American higher education. While the newly created public colleges and universities worked to define the significance of a college education for mass public consumption, Thomas Arkle Clark was busy creating a role for himself, which became the model for other deans of men. He set the standard of expectation for student behavior and conduct on his own campus, but he became a moral compass for young men everywhere. Using his position as Dean of Men at Illinois as a fulcrum to leverage public opinion, he created a larger-than-life persona for himself that other deans could only imitate.

Illinois, like many other public universities in the early decades of the twentieth century, grew by leaps and bounds. No longer the small, middling college on the prairie that Clark had attended as a boy, Illinois added engineering, agriculture, and other colleges to its liberal arts core. This rapid expansion drew many more students. When women began to enroll on a regular basis, the institution became a beacon of intellectual and social opportunity for many young men and women across the state and beyond.

Yet as the campus expanded with greater student enrollment, few administrators and even fewer faculty concerned themselves with the activities of students outside the classroom. Only vaguely aware that students had lives beyond the classroom, few adults on campus took students' extracurricular activities as significant or worthy of their attention. This vacuum between faculty and administrators was filled by deans of men and women. As the first dean of men, Thomas Clark stepped into this void and created a model in his own image of what a dean of men should be.

Because of his increasing administrative role, Clark gradually abandoned his rhetoric classes, but he never stopped teaching. His teaching role simply moved from the classroom to his administrative offices, where the well-worn and infamous "green carpet" covered the floor. Over time, many young men at Illinois would find themselves standing on that carpet, and remained under Clark's dominion until they graduated or left town. Clark clearly believed he had a moral as well as professional obligation to teach young men and to a degree, young women, in his role as dean.[19] He used his interactions with individual students and student groups, such as fraternities, as well as the campus newspaper and eventually radio, to extend his "lessons" across the campus.

As Clark's work and scope of authority became more extensive, he hired several assistants, young men of like mind and temperament, who worked with Clark on various tasks and assignments. The first official appointment in the Dean of Men's office occurred in the 1912–1913 school year when Professor A.R. Seymour was named as Adviser to Foreign Students and reported to Dean Clark. In the 1913–1914 school year, Arthur R. Warnock became the first Assistant Dean of Men. In 1918, Seymour's title changed to Assistant Dean for Foreign Students. In 1919, Warnock left to become Dean of Men at Pennsylvania State College (now University). He was replaced by Horace B. Garman. Seymour became Dean of Foreign Students and Herbert Creek was made Assistant Dean for Foreign Students.[20]

These staffing arrangements continued to evolve as the years and responsibilities rolled on and the Office of the Dean of Men expanded. By 1922, Clark was supported by Robert Gardner Tolman, who was Dean for Freshmen, and Fred Turner became Assistant to the Dean. In 1923, James G. Thomas was added as another Assistant Dean. In 1924, Turner ascended to become Assistant Dean and Thomas became Assistant Dean for Freshmen. By 1925, Roger Hopkins was added to the staff as an Adviser to Student Organizations and Activities.[21]

These administrative changes reflected the rapid expansion of student enrollments at Illinois. But make no mistake—the growth of the Office of Dean of Men at Illinois was a direct reflection of Thomas Arkle Clark's deliberate use of his office to gain more and more administrative control and to expand his power and influence beyond the scope of student discipline and individual advising. Clark's reputation beyond Illinois grew by leaps and bounds. It is hard to believe that Clark's fame was accidental. He published on a regular basis, he spoke on radio programs at Illinois, students who knew of him while on campus also distributed his legend across the Midwest and beyond as alumni. Clark's involvement with Alpha Tau Omega fraternity, a social organization with numerous chapters across the United States, was a ready vehicle for marketing his name and building his reputation. He resigned the position in 1931, the same year as his retirement, on advice from his physician to curtail his extra activities due to "overwork."[22]

Clark's fraternity affiliation was not limited to Alpha Tau Omega, but extended to many others through his association with the National Interfraternity Conference and other national fraternity leaders. One of his most powerful allies was L.G. Balfour, president and chief executive officer of the Balfour Company, the jewelry company that supplied the many fraternities and sororities across the nation with pins, lavaliers, and other hardware. Balfour, known to his friends as "Bally," was a determined entrepreneur who kept up a steady drumbeat of enthusiasm for fraternities and the benefits of fraternal membership, while at the same time fighting ferociously against those individuals who might limit the fraternity on campus.[23]

George Banta, editor of *Banta's Greek Exchange*, was another nationally known fraternity colleague and Clark champion. The *Greek Exchange* was a national magazine that spread the news of the fraternity world across the United States on a monthly basis. Clark was cited frequently in the *Exchange*, often as an authority on any matter related to fraternity life and college in general. *Banta's Greek Exchange* was a powerful medium for correspondence and communication among university administrators, including college and university presidents, fraternity and sorority national officers, alumni, and undergraduate chapters. Closely read for the news of the day in the fraternity community, the *Greek Exchange* supplied news, notes, opinions, and more for a very widespread audience. Clark was featured in the *Exchange* for a number of years, often cited as a campus-based expert on fraternity life.

Clark's reputation was further solidified by his regular appearances at national conventions, in publications, and in many speaking engagements across the country. Clark was seriously and deliberately engaged in furious self-promotion, whether at the National Education Association meetings, where he spoke on the values of fraternity membership as early as 1910, or in his newsy and colloquial columns in the *Greek Exchange* as well as other publications.

Clark wrote extensively, which helped cement and enhance his reputation. His writing also extended his renown far beyond Champaign and Urbana, Illinois. Initially, Clark wrote essays for the consumption and edification of young men on campus. These essays gradually became accounts in the student and local newspapers, carrying such titles as Faith, Gossip, Investments, Leisure, Youth, Responsibility, Learning to Talk, Resolutions, and so on. Some were directed toward character development of the young man and bore titles reflecting those tasks, such as The Joiner, The Boy and His Clothes, The Dance, The Leisure Hour, The Liar, The Prodigal, The Politician, The Week-Ender, When You Know a Man, Father, Christmas, Commonplace, and Excuses.

The proliferation of these treatises was significant. Clark pounded them out on a regular and consistent basis, week in and week out. His commitment to his writing eventually resulted in a total of some 1400 essays with titles such as Doing Without Sleep, Plenty of Meat, Playing Second Fiddle, and The Acid Test. Most of Clark's homilies addressed character-building concerns, whereas others were specific to issues of morality, for example, Saving Oneself and The Art of Lying. Some of the essays were collected into books with titles such as *The Sunday Eight O'clock: Brief Sermons for the Undergraduate* (1916);

The Fraternity and the Undergraduate (1917); and *Discipline and the Derelict: Being a Series of Essays on Those Who Tread the Green Carpet* (1921). Other essays served as radio addresses on the campus radio station or as the source for speeches to various groups and conventions, such as the NADM and the National Education Association.

Through a steady stream of publication, speeches, and public pronouncements and appearances of one sort or the other, Dean Clark became synonymous with campus life and the culture of college in the early part of the twentieth century. Clark became so well known as a public figure that he was even featured in a Quaker Oats advertisement in the *Saturday Evening Post*, arguably one of the most popular magazines in the United States at the time.[24] Clark's photograph appeared in the ad with another, more established American icon of the time, Knute Rockne, the football coach at Notre Dame. Even so, it was Clark's picture that appeared at the top of the ad with Rockne located below. It was an impressive achievement for an orphaned farm boy from Illinois, selling a popular breakfast cereal in a national magazine with the head football coach at Notre Dame.

FRATERNITY AS SOCIAL ORDER

Although a devout disciple of the college fraternity himself, Thomas Clark turned the college fraternity to his own ends as a dean. Shrewdly, Clark saw how fraternities could be a "social leavening agent" on campus and was able to use the fraternity and fraternity membership as a tool for control of student culture and campus order. By controlling the students, who were primarily male, through their organizational affiliation, Clark's job was much easier than if he had attempted to effect change or control with individual students one on one.

Like many other public universities in the early 20th century, enrollments at the University of Illinois rose from 751 in 1894–1895 with 627 men and 124 women to 2181 students by 1903–1904 with 1610 men and 571 women. Social fraternities grew along with increasing enrollments and at Illinois, Clark encouraged their proliferation. By 1930, near the end of Clark's career, the number of fraternities at Illinois had reached 92 chapters.[25] Sororities grew as well, with 33 sororities established by 1930. Illinois, under Thomas Arkle Clark's benevolent direction, became one of the largest fraternity systems in the United States.[26]

What men in fraternities wore, what they studied as academic majors, and especially their campus behavior, had a huge impact on campus through peer influence and control. Rather than addressing one student at a time, Clark was able to influence the fraternities on a chapter-by-chapter basis or through the governing council, the Interfraternity Council.

Ironically, in 1870, two decades prior to Clark's arrival on campus, fraternities had been seen as the antithesis of academic life by many college presidents. Nowhere were fraternities less welcome on campus than at the University of Illinois. Fraternities were strictly prohibited by the first two presidents of the university, John M. Gregory and Selim P. Peabody. Gregory and Peabody saw the fraternity as "anti-intellectual" and against the best interests of the university. Worse, both men worried that fraternities would undermine their authority as presidents amongst the student population, a suspicion that proved to be quite accurate.[27]

The presidential prohibitions against fraternities at Illinois reached such levels that in 1881, all senior students who were planning to graduate were required to sign a statement swearing that they had not joined a fraternity while at the university or risk immediate expulsion.[28] Like most efforts at prohibition, the fraternity ban at Illinois failed miserably. Fraternity members who wished to graduate in 1881 simply lied. It is well documented that at least two social fraternities—Delta Tau Delta and Sigma Chi—existed as sub rosa organizations and continued to operate in secret for several years until a court order forced the university to vacate the ban.[29]

As campus enrollments swelled in the post-bellum period of the late nineteenth century, students effected changes by waging a war of attrition. College presidents like Gregory and Peabody at Illinois tried and failed to maintain the traditions of the "old time college." Student opinion and attitudes at the end of the nineteenth century would not tolerate such heavy-handed tactics. Students demanded a more active campus social life and a greater say in their affairs. Intercollegiate athletics rose in popularity, especially college baseball, which predated football and basketball. Fierce rivalries grew between colleges in close proximity and student loyalty to their schools soared. Fraternity and sorority memberships increased, as did interest and membership in other student organizations. Students found themselves united by their memberships in student groups and organizations, and in their support of their athletic teams. They also found common voice in protesting, sometimes violently, against college rules and regulations they disliked. Military drill, a mandatory requirement of all male students at land-grant colleges and universities, was a favorite target for student protest. Thomas Clark recalled a riot at Illinois during his undergraduate days when he saw "a cannon backed into the sluggish stream that flowed thru the campus to show disapproval of compulsory military drill."[30]

Many of the men who became deans of men had, like Clark, been students during this period of turmoil and transition. The deans straddled two eras of college life, the end of the "old time" college of eighteenth and nineteenth century America and the new, emerging institutions of the twentieth century. From the 1880s onward, students applied constant pressure to colleges and universities. Through overt and subtle tactics alike, students slowly wrestled more and more authority away from presidents and trustees to reshape campus culture. Students became social "change agents" according to DiMartini.[31]

In the midst of such an evolution, no one could be quite sure where colleges and universities were headed. However, it was clear that they were becoming uniquely American institutions. In their effort to work with, not against, such changes, the deans of men viewed student organizations as assets to their management style, not impediments.

By 1910, 30 years after the ill-fated fraternity ban at Illinois, the changes were obvious. Speakers at the National Education Association meeting that year included W.H.P. Faunce, president of Brown University, who praised fraternities in his speech, "The Relation of the College Faculty to Fraternities." Ralph Jones of the University of Maine discussed a survey he had conducted on the merits of the college fraternity. His survey demonstrated, Jones argued, that fraternities "... will do more to bring about radical improvement in the college work of undergraduates than the entire faculties of the [173] colleges [where] fraternities exist can accomplish in five times that period."[32] Another participant, Dean Clark of Illinois, praised the social fraternity as a "leavening agent" on campus which set standards and controlled students. "I have favored fraternities, and other social organizations... because I have found them of the greatest help to me in controlling and directing student activities."[33]

In times when student enrollments were growing and student dissent was never too far from the surface of campus life, Clark and other deans showed remarkable poise and intuitive sense of their campuses. One of the best means for them to control the student populations was through control or at least manipulation of the student organizations, and few organizations at the time mattered more than the college social fraternity.

Clark was an early member of the Gamma Zeta chapter of Alpha Tau Omega fraternity at Illinois. Initiated in 1895 at age 33, early photographs show Clark, with a large and distinctive cookie duster mustache, at the ATO formal dinner in 1901 on campus. His loyalty to the fraternity extended far beyond his undergraduate experiences into his adult life. Clark was elected to be the Worthy Grand Chief of Alpha Tau Omega, "the highest elected officer in the fraternity from 1918 to 1923 and again from 1929 to 1931," a job he approached with great relish and zeal.[34] He traveled the country as much as he could, speaking on college campuses and to alumni groups as often as he could. The position combined two things Clark valued highly, his fraternity and travel, so it was a very pleasant year indeed for the Dean.

As the number of fraternities at Illinois grew, almost exponentially, over time, Clark's influence and stature in the fraternity world also grew. Not only was he seen as the "dean of deans of men" but quite possibly the most influential dean in the fraternity world as well. He certainly made his appreciation for the college fraternity known and encouraged the growth of the fraternity system on his own campus. As noted by the careful recording of fraternity memberships among the deans of men at the early NADM meetings, Clark and the other deans of men shared a common commitment to the "organized," as Horowitz has called them.[35]

What is clear about Thomas Arkle Clark and his role as Dean of Men at Illinois is that Clark valued social control above all else. He was determined to be successful in assuming responsibility for the behavior of Illinois students and demanded a fairly high level of responsibility, some might argue obsequiousness, in return. In a more modern world, where student rights and responsibilities seem to limit such intrusions, Clark's methods might appear to be heavy handed and overly paternal. But in the early 1900s, Clark's manipulations of young men under his charge was rarely challenged. He deliberately sought out and exercised as much control as he could manage as an administrator and used it to establish a powerful role for himself on campus.

In particular, Clark's manipulation of the fraternity members and fraternity chapters on the Illinois campus was very successful. In many respects, it was a mutually satisfactory relationship. Through Dean Clark, the fraternity men had a staunch and abiding ally against the antifraternity forces such as faculty, other administrators, women, religious groups, or even non-fraternity students. The "unorganized" nonfraternity men must have harbored some jealousy of Clark's overt attention to fraternity men, to say the least.

Clark was able to maintain order and control among most of the students most of the time. If Clark perceived his role as the primary administrator in charge of student behavior, he was quite successful at it. For a poor, orphaned boy from a farm background in the rural prairies of downstate Illinois, Clark's success was heady stuff indeed. It was even more satisfying that the University of Illinois was the flagship university of the state and his alma mater.

However, Clark's manipulations and demands were not without drawbacks. There were students who came down on the wrong side of Clark's sense of values. For them, there was little recourse. Dean Clark may have been beneficent and fatherly in his outward demeanor. His dinner speeches, public addresses, and his radio addresses to the campus community were warm, personal homilies to personal virtue, pluck, and individualism. But for those who crossed his path or violated university rules, Clark was a formidable enemy.

Clark mounted a one-man crusade against bootleg liquor, wanton women, and gambling, all of which were readily available in Champaign and Urbana, Illinois, in the 1920s. A dues-paying member of the local temperance league, Clark used his fraternity contacts as a source of information to uncover some of the illegal activities on campus. In the flush of a quiet confrontation in the Dean's office, faced with expulsion or at the least, a disciplinary sanction of some sort, many a young man capitulated and blurted out information to the Dean.

Clark eventually wrote up many of his disciplinary encounters in a book, notably titled, *Discipline and the Derelict* (1921), ominously subtitled, *Being a Series of Essays on Some of Those Who Tread the Green Carpet*, a subtle reference to the infamous green-colored carpet in Clark's office. He also gave radio addresses—the Sunday Eight O'clock—many of which were later published in the *Daily Illini* student newspaper. Clark's print presence, both on campus and then nationally, was impressive. From the moment students first set foot on campus until graduation, Dean Clark was omnipresent. A glimpse of Clark's intent to educate more than punish is found in the book version of his radio essays on virtues, manners, behavior, and the like, *The Sunday Eight O'clock: Brief Sermons for the Undergraduate* (1916). Clark also syndicated some of his 1400 essays to newspapers and magazines.[36] Clark drew on his extensive knowledge of the issues confronting young men to wax philosophically on how to study, how to dress, how to behave in pubic, how to manage money, and so on. For many young men away from home for the first time, these lessons were the quintessential form of *in loco parentis*, reminders of parental supervision at home that were absent on campus.

DISCIPLINE AND THE DEAN

Clark recognized that the transition from home to college was a difficult task for many students. Clark's homilies were a quaint but valuable lesson in how to succeed in college for those alone and buffeted by many changes and challenges in an uncertain and new environment. He was there in his addresses to the freshmen, he was there in notes and messages left in student mailboxes, he was there in the student paper, and later on the radio. His message was pervasive and unending—follow the rules, live a clean life, go to class, honor authority. Failure to abide by these strictures, especially for those who got caught, typically meant a college career cut short by expulsion from the University of Illinois. Beginning in 1911, all new students were given a handbook called "Facts for Freshmen," loaded with rules and regulations, advice and admonitions, beginning with the first arrival in the city of Champaign,

> You can come into Champaign or Urbana by the Illinois Central, the Big Four, or by the Wabash railroads, or by the Illinois Traction System. Whichever way you come a local electric car will land you at the University grounds within a few minutes. You will be met at or on the train in the Fall by all sorts of commissaries or representatives, each of whom will offer to conduct you about, and will at the same time solicit your patronage of is boarding club or lodging house, or other particular pet scheme. Go slowly.[37]

In pursuing his control of the Illinois campus, Clark employed a far-ranging network of informants. He used the local police, as might be expected, but also took advantage of the services of other less obvious sources, such as coaches, faculty, church members, and townspeople, who saw themselves as doing a good service by reporting on student behavior. Unsolicited messages and information from the community fed the information flow in the Dean's Office. A typewritten note sent to Clark on November 11, 1930, simply stated, "Dear Sir: This is to advise you that Mr. Jack L. Sachs and Miss Frances Knight did sleep together in the same bed in an apartment in Champaign in the evening of November 2nd until next morning. Please investigate. Very truly yours, Citizen."[38]

Clark also relied on the services of the Pinkerton detective agency, who patrolled the railroads and railroad yards looking for tramps riding the rails and vandals.[39] Because the railroad tracks often paralleled the seamy side of town, the Pinkerton detectives kept a close eye on roadhouses and prostitution, and they willingly passed on information to Clark for his use in intercepting wayward students who might frequent the same dens of iniquity.

Clark was a good source of information himself. A letter from Harold Baker, U.S. Attorney for the Eastern District of Illinois, to E.C. Yellowley, the Federal Prohibition Administrator for the State of Illinois, dated March 5, 1928, reported that

> Dean Thomas Arkle Clark, Dean of Men at the University of Illinois called me by long distance and reported that violations of the Prohibition Act were again numerous in that vicinity.... You will recall that we received splendid cooperation from the University authorities in the cases made there last Fall and last Spring. I believe you should endeavor to aid these men in their work as much as possible and would respectfully suggest that, as soon as possible, you detail certain agents to investigate in that vicinity."[40]

Two weeks later, Baker wrote to Yellowley again. In this letter, dated March 16, 1928, Baker reported on specific locations of corruption. In a face-to-face meeting with Clark in Champaign, Baker recorded what he called a partial list of locations where the Prohibition Act had been violated. Baker's letter detailed the locations as:

> Bill Gross Garage-Rear Princess Theatre- Urbana;
> Becker's Drug Store-Urbana;
> ____Daily on Elm Street just west of Police Station-Urbana;
> a gambling and drinking den over Bullock's Drug Store in Champaign.
>
> Dean Clark assured me that if your men would come to him he would be able to give them more and complete (sic) information by the time they arrive. He suggested that someone similar to [Eliot] Ness be sent to make the investigations for he was well pleased with the methods employed by Ness and the results obtained. He [Clark] suggests that should your agents go to Urbana that they stay at the Grenada which is located at 1004 South Fourth Street in Champaign, where there are located certain students who have certain information relative to these places and who can probably divulge it to your investigators.[41]

Eliot Ness, still early in his career with the Treasury Department and not yet pursuing Al Capone, had worked the year before, 1927, with Dean Clark in an effort to break up illegal bootlegging in Champaign and Urbana. At the time, Ness was working with the Bureau of Prohibition in Chicago and the trip to Champaign was a relatively short drive or train trip.[42]

In his second letter, Baker also indicated that two local pastors could corroborate Clark's concerns about the Prohibition violations in the Champaign-Urbana area. From this disjointed but interrelated net of communication and gossip, it seemed clear to many students that Dean Clark was everywhere and often ahead of them. Rumors were rampant about the extent of Clark's exploits in catching miscreants. One rumor circulated that Clark had climbed down a chimney to break up an illegal party, that he could read on a man's face whether he was telling the truth or a lie, and that if caught, he was merciless to the guilty. Some of these rumors were true, others were patently false. Clark never climbed down a chimney but he probably relished the story himself. Depending on the student, and the conditions under which they met, students either loved Clark or despised him.

TWO SIDES OF THE COIN

Two incidents of student encounters with Clark illuminate the substance of the many Clark myths, and demonstrate that many of the myths had a basis in fact. William O'Dell was from LaGrange, Illinois, and entered the university in 1926. O'Dell was to have graduated in 1930 but was expelled by Dean Clark, which delayed his graduation for one year. As O'Dell described it,

> There was a University "no car" rule at that time and undergraduates could not drive cars unless they were employed [off campus]. I got my parents permission to take the family car from LaGrange down to Champaign for a big Spring weekend, a dance weekend. So I had the car in violation of their no car rule. And this made travel with my girlfriend from one fraternity to another easier, and we could go to many different fraternity parties during the weekend.
>
> I was driving down Green Street [a major road] with 3 or 4 people, en route to Chicago, or La Grange, to return my car and then take the train back to Champaign on Monday. There was a note in my mailbox from Dean Thomas Arkle Clark asking me to come to his office at 11:30 the next morning. This would strike terror into anyone's heart. And so I showed up. He must have been 120 years old [Clark was 68].
>
> My interview with him was maybe 40 seconds long, and the message was that I was being expelled from the University for the balance of the semester for violating the no car rule. He named the people in the car, what direction I was going down Green Street. The rumor, of course, around the campus was that Thomas Arkle Clark had a spy

> ring, and that had never been documented, but I don't know that it had ever been completely denied either. It was my last semester in school, in May, two weeks, maybe three weeks away from my degree. It cost my father another semester's education. Very unfair, in retrospect, a horrible penalty for such a relatively minor infraction, in my opinion!
>
> In any event, Thomas Arkle Clark was a feared dean. I had tremendous respect for him, but that was the rule and regulation that I do remember. I never saw him again but he was a very stern individual."[43]

Clark's antipathy toward the automobile on campus was legend in and of itself. Like many campus leaders, Clark saw the automobile as the primary instigator of moral corruption. In a 1924 letter to parents, Clark railed against the car.

> The automobile is a waster of time and money. It encourages loafing and the taking of frequent and unnecessary trips out of town to the neglect of the students' regular work. The student with a car is likely to be a poor student because he cannot resist the temptation of the car, in spite of the fact that he usually promises the home folks that he will use the car only at week ends.
>
> There is moral danger in the car. Whatever of drinking and stealing, and sexual immorality exists among college students is largely in connection with an automobile. The passion for driving seems often to stimulate other passions and unconventionalities and actual immorality often results.[44]

By the time O'Dell had reached campus and committed his transgression in 1930, Clark's position on automobiles had not changed an iota and he had no mercy for those few students who violated the rule.

Another student, Royal Bartlett, who graduated in 1931, recalled Dean Clark with more positive regard. Entering the University of Illinois in 1927, Bartlett described the challenges of trying to go to college during the Depression years. Any extra expense could cause problems. Bartlett's father had passed away just before he went to college so he altered his plans and intended to get a job to help support his family. But his mother insisted that he go to college anyway. Bartlett borrowed money from a hometown bank and lived as frugally as possible.[45]

Bartlett had little direct experience with Dean Clark except for a single incident when some cleaning went missing. In his words, Barlett recalled that when clothing was sent out from his boarding house to a local cleaners, a delivery man from the cleaners would return it to the student's building in the foyer. However, Bartlett lost a suit that never appeared. Bartlett noted that there was another boarding house just like his across the alley so he went there to see if his suit had been delivered there by mistake. But he had no luck finding his lost suit there either. When he complained to the cleaners, they responded that according to their records, the suit had been delivered to Bartlett's address and was no longer their responsibility.

In Bartlett's words, "A suit of clothes at that time was probably $20 and that was a lot of money to have, so I went to see Thomas Arkle Clark to see what the University would do with this dry cleaner. And so he made a phone call or two and the next thing you know, I had $25 to buy a new suit. That was my only experience with him [Clark]. But he was well liked."[46]

These two student accounts are quite contradictory in terms of the students' experiences of Dean Clark. But they accurately describe the very different aspects of Clark's campus persona. In O'Dell's case, his experience with Clark was framed by his own acknowledged violations of university regulations. His fear of Clark was compounded, to some extent, by his own guilt. In Bartlett's case, a poor young man struggling to get through school has some bad luck and his encounter with Dean Clark is a pleasant one—Clark interceded on his behalf, he is compensated for his lost clothes and feels good about Clark, the university, and his campus experiences.

Clark was a compassionate man who cared about students on both a personal and professional level. But he would not countenance violations of university rules, most of which he had created or at least enforced. His role on campus was that of a benevolent dictator who ruled the lives of the young men under his authority. Clark's absolute dedication to his sense of moral behavior guided his years as dean and his interactions with students. In O'Dell's case, Clark brought the swift sword of justice down on the young man's career at Illinois. For Bartlett, Clark was caring and compensatory; he used his influence in the local business community to get the young man the money to buy a new suit.

Clark valued honesty and what he called acting in a "manly" way above almost all else. These behaviors were windows to a young man's soul, in Clark's view. He could not tolerate the underhanded or sneaky behavior often found among young men who, like O'Dell, even when caught, might try to hide or minimize their guilt. To those miscreants, Clark's notion of campus justice was swift and fast.

Expulsion was often the response, although the truly repentant might be allowed to return.

In Clark's own words, the typical pattern becomes clear as does his thinking.

> Two years ago, I had another experience with a young fellow caught in a really serious college escapade, which strengthened materially my faith in human nature. It was a situation in which the boy knew that if he told the truth, he would be permanently dismissed from the college. I knew all the details of the case but this fact, he was not aware of.
>
> In spite of the penalty which he knew would be inflicted, and ignorant of what I already knew, he told our committee as frank and straightforward a story as I ever heard, and though his father is a man of wide influence in the community in which he lives, the boy accepted his punishment in a thoroughly manly fashion and left me with the most friendly feeling. It gave me the greatest satisfaction a few months ago to write him that because of his truthfulness and the manly way in which he had received his punishment, our Council had reconsidered him to return to the University the next Fall—an action which had been taken in reference to no other similar offender in ten years.[47]

These moments of epiphany among young men were the moments Clark hoped for as a dean of men. Through such experiences, he reassured himself and others that these difficult experiences, like his own childhood and adolescence, were the building blocks of moral character and high virtue. Without them, Clark believed, young men might wallow in self-pity, ingratitude, and selfishness for a long time. The confrontations with strict rules and regulations at the university were moments of truth for Clark. Unlike conciliatory parents who might indulge a young man's fancy, Clark was unbending or at least appeared to be so until there was some form of contrition expressed.

To many male students, Clark was a man to be, at the least, respected, and for those Clark referred to as "the derelicts," he was feared. But he was far from the ogre O'Dell recalls. Clark was just as often involved in kindly acts, as noted by Bartlett. . Other men recounted similar acts of generosity and kindness throughout Clark's tenure as dean of men.

THE END OF A CAREER

Clark retired in August 1931. The year before, on May 11, 1930, a celebration of Dean Clark's sixty-eighth birthday was attended by over 300 people in Urbana. Seven invited speakers celebrated Clark's influence on young men at the University of Illinois over the years. Over 300 letters from across the country arrived to honor him. Later that summer, Clark attempted to retire but he was convinced to stay one additional year at the request of the new Illinois president, Harry W. Chase, primarily to maintain stability in the new president's administration.

In November 1931, Clark had abdominal surgery for removal of a tumor at Presbyterian Hospital in Chicago. He recuperated through the months of December and into January 1932. In January, he and Alice traveled to Arizona for recuperation for three weeks in a milder climate. However, the cancer returned. On July 15, Clark slipped into a coma from which he never recovered. Thomas Arkle Clark, age 70, died three days later on July18, 1932. The official cause of death was listed as cancer.

A large and respectful funeral was held in Urbana as Clark was buried with friends and admirers looking on. The proceedings were supervised by his Assistant Dean and successor, Fred Turner. Alice Broaddus Clark, his wife of 36 years oversaw the arrangements and attended the funeral. Clark was buried in Champaign, several blocks from the campus where he spent the majority of his life. Out of respect for his sartorial reputation and the season, Clark was buried in his famous white flannel summer suit.

His obituary in the *New York Times* read, "In his last year in office, [1931] Clark saw an average of 352 students per day, most for loans due to the Depression. Clark was also credited with the development of a 'campus spy system' for which he was perennially criticized by students. His defenders said the spy system was a myth, that the dean merely made use of information elicited in chats with students who were due to be disciplined." Later in the same obituary, Clark was reputed to be near superhuman in his zeal to find errant students. "In his younger days, when drinking was merely a violation of university rules, the dean would drop down fraternity fireplace chimneys to surprise a group of young men over steins of beer."[48] Some myths were meant to live in perpetuity.

The greatest set of accolades for Clark and his many years of service were accumulated in anticipation of his resignation as Dean of Men in 1931. Tributes were offered from a wide ranges of sources, from *Banta's Greek Exchange* and the Alpha Tau Omega magazine, to the *Rotarian*, the national magazine of the service organization, the Rotary Club. Alumni of the university, deans and presidents from other universities and colleges, and others sent letters of appreciation and gratitude to Clark for his many years of leadership and service.

At the time of his death, Thomas Arkle Clark had made a significant contribution to American higher education and an even greater contribution to the rapidly emerging profession of dean of men. The out-of-class activities and behavior of young men on college campuses across the country had become the responsibility of a select group of administrators, deans of men. As institutions expanded both in the number of colleges and universities as well as the sheer number of students on campus, the role of administrators and faculty directly involved with student life outside the classroom receded. The early decision of Charles Eliot at Harvard to create a dean for faculty and a dean for students created a model for university administration that was copied nationwide. However, it appears to a large degree to have absolved many adults on campus from direct contact with students outside the classroom. Consequently, it was left to the deans of men to step into the void and create a role for themselves that they had to define by trial and error.

As the first dean of men, Thomas Arkle Clark used his office for a variety of personal and professional opportunities. Clark took his office seriously and performed his duties in great earnest. He adopted a fairly stern approach to the escapades of young men at Illinois. Automobiles, alcohol, and cheating were cardinal sins in his view and typically grounds for expulsion. However, Clark often was willing to forgive the repentant and students who were discharged were often able to return to complete their studies if they acknowledged the errors of their ways. Clark also had a soft spot for young men who were limited in their means and opportunities. Many of the young men who attended the University of Illinois in the late 1800s and early 1900s were from rural, farm families or small towns. Clark felt a strong affinity for young men trying to better themselves as he had himself and granted them extra consideration when he could.

Clark presented a strong moral compass for young men and even women. Led by his own Presbyterian conservatism, Clark was a strong advocate for the temperance movement. He collaborated with the Pinkerton detectives hired by the railroads that crossed through Champaign-Urbana to keep track of the roadhouses and gin joints in town. He managed to stay one step ahead of most of the students through this informal network of information.

More formally, Clark and the Office of Dean of Men published a booklet outlining the rules and regulations for student behavior. Student were expected to know and read the booklet and were held accountable for their actions. Student handbooks and codes for student life are still a part of the contemporary campus, although more riddled with legal language and wording than the early versions found at Illinois and other campuses. Clark also used campus media, initially the campus newspaper and later, radio broadcasts, to disseminate his message. His Sunday Eight O'clock lectures became essays on moral behavior, conduct, and how to conduct oneself in the world directed primarily at students, but the appeal of the clear, unvarnished directives became popular with an audience off campus as well.

At the same time, Clark basked in the glow of a wider audience for his paternalism, which found commercial success as well. Clark traveled extensively to lecture and promote himself and to a lesser degree, his university. As noted earlier, he became so well known nationally that he was featured in a Quaker Oats advertisement along with Knute Rockne, the Notre Dame football coach. Few other academics could boast of such fame at the time nor have many since that time. Neatly turned out in his favorite white flannel suits, Clark, by the time of his retirement in 1931, had achieved national stature and standing; no small achievement for an orphaned boy from an Illinois farm.

3

The Paternalists

Image 2 Scott Goodnight. Courtesy of the University of Wisconsin-Madison Archives.

By the middle of the fall term in 1919, Scott Goodnight, Dean of Men at the University of Wisconsin, was quite frustrated. He had tried communicating with the other deans of men in the Big Ten conference via telephone or by telegram but he found these contacts very limited. Goodnight wanted to sit across the table, face to face with his colleagues, and hammer out mutual problems. Goodnight's primary concern was disciplinary issues associated with the athletic events between the schools. In many cases, he thought the problems were beginning to get out of hand. A frank discussion among the deans might help eliminate some of the problems.

So, in his own words, Goodnight "wired Minnesota, Iowa, Illinois, Indiana and Michigan to come on over." As a result, a small group of deans of men, responding to Goodnight's invitation, traveled to Madison, Wisconsin, in January of 1919. The minutes of the first meeting reported that

> the following topics were considered at one or various times: (1) Attitude toward major activities, (2) Fraternity initiations, (3) Fraternity finances, (4) The Warner system of fraternity management, (5) Relations of students and landlords, (6) Classroom attendance and scholarship, (7) Credit for military work, (8) Student self-government.[1]

Professor Louis Strauss, representing the Faculty Senate Student Affairs Committee at the University of Michigan, kept the minutes of the meeting. He initially sent Goodnight a set of intentionally lighthearted notes in fun, a set he intended only for Goodnight's enjoyment. However, Goodnight urged his colleague to send the minutes out as written. So the surviving record of the first meeting of deans of men states that Scoot Goodnight was elected chairman of the group when "...in the midst of a heated argument it was discovered that this formality [appointment of a chairman] had been overlooked and also that the bottom of the box of cigars generously provided by the host had begun to heave into sight."[2] Strauss teased Goodnight about his driving skills, noting that there were "Curbstones damaged—1, Persons run down—None [and] Bawled out by police for reckless driving, etc.—4."[3] The meeting ended with a "smoker" hosted by the Wisconsin faculty at the Kappa Sigma fraternity house.

Despite Professor Strauss' humorous account of the meetings, much was accomplished over those several days in January 1919. Goodnight and five of his fellow deans, including Thomas Clark of Illinois, met, conversed, and conferred on the role of their positions as deans of men. The deans found such value in the session that they agreed to meet again the following year at the University of Illinois. It was at this second meeting in 1920 that the assembled deans of men declared themselves to be a new professional association, the National Association of Deans of Men.

Goodnight himself had little sense of the future direction of the group of deans. He was far more concerned with the immediacy of his own campus concerns and was determined to expedite those matters as quickly as possible. Efficiency and expediency were critical watchwords in the Goodnight lexicon. Prior to becoming dean of men, Goodnight was already a busy man. He was an associate professor in the German department at Wisconsin, and in the summers, assumed the role of Director, and later Dean, of the Summer Sessions.

Initially, Goodnight saw his role as Dean of Summer Sessions as a far more important responsibility than his duties as Dean of Men. Bringing students to campus in Madison, Wisconsin, on the shores of Lake Mendota was no small feat in the early 1900s. As the program grew, the summer enrollments often involved not just students, but families, including wives or husbands along with small children. By the summer of 1915, the summer session programs and offerings had grown so large that a virtual "tent city" was created, accommodating several thousands of people.[4] By comparison, the job of Dean of Men, which Goodnight assumed in 1916, involved less planning and fewer accommodations. Over time, however, Goodnight gave up his teaching responsibilities to become a full-time dean of men.

He was born in Holton, Kansas, on January 16, 1875, the son of a minister in the Christian Church. It was the custom of the Christian Church that no minister would serve a single church for more than a year at a time. As a result, the Goodnight family moved around the Midwestern states as Goodnight was growing up, finally settling in Illinois. Goodnight attended Eureka College in Eureka, Illinois, a small liberal arts school affiliated with the Christian Church, where he earned a bachelor of arts degree in German in 1899. (Eureka College would later gain greater attention as the alma mater of President Ronald Reagan.)

After his graduation from Eureka in 1899, Goodnight spent a year in Europe, taking classes at the University of Leipzig and the University of Neuchatel in Switzerland.[5] He returned to Eureka the following year and, in his own words, "taught all the French and German classes at Eureka. I was in the classroom 25 hours a week and didn't even have time to read the newspapers."[6] While teaching as Professor of French and German, Goodnight earned a masters degree from Eureka in 1901. He also married his college sweetheart, Gertrude Hamilton, a fellow graduate of Eureka, in the same year. The Goodnights moved to the University of Wisconsin in 1902, where Scott Goodnight completed his doctorate in German. He gained a position as a graduate teaching instructor in German, attributing his hiring as a graduate assistant, in part, to his athletic prowess at Eureka.

> They told me there was a job for a German instructor at the University of Wisconsin. Although there were 30 applicants, several with academic qualifications considerably beyond mine, I got the job on my football record.
>
> Professor Hohlfeld had recently come from Germany to head the Wisconsin German department. A little man, he was terrified by 'the American wild West,' especially those big students in engineering who wore corduroy pants, let their whiskers grow, and did other things that Hohlfeld didn't understand.
>
> Hohlfeld wanted a man to teach German to the engineers and was more concerned with his size and toughness than with his erudition in German. I measured up in size, all right, and when he found out I had also played football—which he considered a wild and rough American game—he hired me with no more questions asked.[7]

Goodnight was employed as an Instructor at Wisconsin from 1901 through 1907. In 1907, he was promoted to Assistant Professor. He became Assistant Director of Summer Sessions in 1911 and was promoted to Associate Professor of German in 1912. Over time, the Goodnights had two children, Scott Jr. and Eleanor.[8] With a growing family, Goodnight worked his way up the academic ladder and in 1912 earned a promotion to Director of Summer Sessions.

Goodnight was appointed to be Dean of Men in 1916, which according to most accounts, made him only the fourth person to hold the office of dean of men nationally. Goodnight served as chairman of the Committee on Student Life and Interests, a committee created through a faculty report on student activities in 1914. The report concluded that greater administrative oversight of student extracurricular pursuits was necessary and asserted the oversight of the faculty through the committee. Over the next two years, the Committee determined that a Dean of Men should be appointed from among the faculty. In 1916, quite naturally, Goodnight moved from chairman of the committee to an appointment as Dean of Men through the action of President Van Hise. During World War I, professors of German were more than a little expendable, so the move to become Dean of Men was more than just a matter of convenience for Goodnight. It provided him with greater job stability and a new line of work.[9]

The appointment of a Dean of Men seemed long overdue at Wisconsin. The appointment of a "Dean of the Woman's Department" was first suggested in the President's Report of 1890 when President Adams recommended that a person "whose education, tact, discretion, and wisdom would recommend themselves to universal favor."[10] Adams left it to his successor, Van Hise, to officially appoint the first Dean of Women, Lois Kimball Mathews, in 1911. A historian by training, Mathews had followed her mentor, Frederick Jackson Turner, to Wisconsin and held a faculty position in the History department as well.[11]

A woman of strong opinions and great intellect, Mathews set a high standard as a dean. During her time at Wisconsin, Mathews wrote a book, *The Dean of Women*, published in 1915. Mathews' book codified the work of deans of women, a book that still crackles with the clear, sharp prose of a woman ahead of her time. Mathews retired in 1916 to marry Judge Rosenberry, a state magistrate. She was replaced by Louise Nardin who complemented Goodnight as Dean of Women and Assistant Director of the Committee on Student Life and Interests for a number of years.

When Scott Goodnight invited the other deans of men from the Big Ten athletic conference to meet in Wisconsin in 1919, World War I was in decline and peace talks were underway in Paris. At the time, Goodnight had been Dean of Men for three years. He was still feeling his way, as he readily acknowledged, and welcomed the guidance and support of his peers, especially Thomas Arkle Clark of Illinois, to define the role of the dean of men.

POST-WAR STUDENT CULTURE

After World War I ended in 1918, young men returned to college campuses in larger numbers than ever before. At the same time, young women began to arrive on college campuses in large numbers, a trend that initiated an unprecedented wave of coeducation. By the middle of the 1920s, the number of women on college campuses nationally had climbed to an unprecedented critical mass. Women accounted for 47 percent of the total undergraduate enrollment in 1925.[12] Although a primary focus of the early deans of men such as Goodnight, Clark and others, was to keep young men in check, the infusion of women onto the campuses of public colleges and universities created a new type of student culture, coeducational, that no one in higher education had ever experienced in such high numbers before.

Not only were the numbers of students greater than ever before but the nature of the students had changed. The young people who came to college in the 1920s were marked by the devastation of World War I; the horrors of mangled bodies and death by invisible gases changed youthful innocence and world views forever. Mass destruction by unseen enemies sucked away the honor and prestige of war. Instead of honorable combat with an enemy, men were killed by the hundreds as they slept in trenches. The nihilism that grew out of massive death and destruction changed even the attitudes of those who never left home. Students took on a new world view that engendered a new culture on campus, a "youth culture" as defined by *Life* magazine in the 1920s.The attitude of many of these new students was to live for today as World War I had revealed the emptiness of living for tomorrow. The convergence of a large number of students on campuses, away from parents and too many to be scrutinized or chaperoned at all times combined with a devil-may-care attitude created a unique culture on the college campuses. It was the age of bootleg liquor, flappers, and the automobile, all inducements to live fast and furiously for tomorrow "we may die."

In a radio address in 1923, Goodnight initially made light of the changes on campus, acknowledging the "fast times" in vogue during the 1920s with the following ditty:

> The genus flapper appeared.
> "Mary had a little skirt,
> So short, so light, so airy.
> It never showed a speck of dirt,
> But it surely did show Mary."
>
> Mary's older sister followed right in Mary's footsteps, however. And then her mother, and soon her grandmother. Shakespeare names the seven ages of men, from babyhood to old age. An observant paragrapher now has named the seven ages of modern woman as follows: 1. babe, 2. child, 3. little girl, 4. young woman, 5. young woman, 6. young woman, 7. young woman. Boys drove fast, gambled, danced, procured bootleg, and drank.[13]

In a later, more reflective moment at the National Association of Deans of Men meeting in 1931, Goodnight took a more serious tone. He explained that the deans of men who were new to their profession and still seeking a consistent definition of their responsibilities, found themselves in uncertain times. Scott Goodnight recalled the mood on campus in the early twenties as erratic.

> The institution of deans of men was really new in this country. There were very few of us, and we did not know of others. We [the deans

> of men] were trying to educate each other as effectively as we could. [What was important was] the smallness of the circle and the intimacy with which we laid our hearts bare to each other... we had been through the great demoralization of the war [WWI] and we were trying to help effect reconstruction in the universities.[14]

The deans faced a daunting challenge. Not only were they unclear as to their roles on campus, but the student population was an ever-moving target. Later in his 1922 radio speech, after the characterization of Mary and her dress, Goodnight characterized the student culture on the campus as follows:

> The World War brought youth into the foreground of public attention. Both old and young bestirred themselves in those exciting days, but it was young men who manned every branch of the service, and it was young women who bore the brunt of the Red Cross work over seas and who, at home, sprang into the breaches... left by the young men in many departments of business and industry.
>
> With the end of the war and the homecoming of the legions from over seas, there was a revolt of youth against a meek return to the subordinate position in the affairs of life which youth had held prior to 1917. Escaped from the repression of military discipline, young men sought to throw off all restraint. And the girls at home, who had patriotically worked and skimped... likewise set up a loud clamor for their share in the thrills and pleasures which had so long been denied. A bountiful supply of money and a spending orgy in the nation at large lent momentum to this movement and there undoubtedly was, for a time, an abnormal amount of recklessness and unseemly behavior on the part of young folks...
>
> Youth... get the credit for the recklessness of the post-war period of whoopee.... It became particularly intriguing to represent college students as low grade morons with flapping sox, fur coats, rickety Fords and not an idea in their heads beyond necking, booze, and sex literature.... [However] in my opinion, the students of to-day are distinctly superior to those of two decades ago, in both scholastic achievement and conduct.[15]

It is worth noting that Goodnight's attitude toward the vast majority of students, especially on the campus of the University of Wisconsin, never wavered from this position. Though he acknowledged that there were a few miscreants within the crowd, he steadfastly defended the quality and wholesomeness of the students on his campus.

THE ORGANIZATION MAN

Goodnight saw his role as Dean of Men to be one of maintaining order and morality on campus, a philosophy similar to that of Thomas Arkle Clark at Illinois. As deans of men, both Goodnight and Clark asserted a conservative moral code by which they regulated the lives of young men on their respective campuses. Not surprisingly, both men came from fairly conservative and religiously influenced backgrounds. Clark's boyhood growing up as an orphan in a hardscrabble farm family was his prize story. He saw himself as a self-made man who had made good due in no small part to his own self-reliant personality. He used his own example as a guide for the young men he advised and counseled. In no small measure, Clark's dedication and deep sense of appreciation for his alma mater, the University of Illinois, where an orphaned farm boy could become a professor of rhetoric and dean of men, was a key part of his philosophy and practice as a dean. As dean, Clark had little patience for those young men who did not see the value and benefit of a college education as he did. For those who tried hard and made an effort, he was tireless in his dedication to their welfare.

Goodnight, too, was the product of a Midwestern family. Raised on a steady diet of conservatism and piety as the son of a minister, Goodnight valued the rural, small town environments of Kansas, Illinois, and Wisconsin. Because the Goodnight family relocated frequently, Scott valued stability and consistency more than most. In part, his long tenure at Wisconsin reflected his commitment to the institution but also his reluctance to uproot his own family.

No stranger to the hard work of rural life, Goodnight developed a strong moral code and strenuous work ethic as a child, which framed most of his adult life. He attended Eureka College, a small liberal arts college affiliated with the Christian Church. His college years were shaped by the same church he had grown up in as a young man. If Goodnight experienced the freedom of college as a young man, it was only within the tight strictures of a small, religiously oriented campus. Goodnight was never far from religion and the strong moral strictures of a ministerial upbringing, even in college. Reflecting on his life, Goodnight noted that he was educated in a "small denominational college and the majority of my best friends of undergraduate days are either ministers or missionaries."[16]

Goodnight's German ancestry was amplified professionally during his time as a professor of German at Wisconsin and later, as Dean, added to his reverence for moral authority vested in social institutions and in institutional officers. In a loose connection to the bureaucratic theories of another German, Max Weber, Goodnight was a true man of the organization. As the University of Wisconsin grew over time, Goodnight's view of institutional responsibility for governing the lives of students grew stronger. The Dean of Men's office maintained much the same attitude and moral high ground in the 1940s as it had in the 1920s despite many social changes over time. Goodnight served as an unwavering moral compass, never faltering in his sense of right and wrong or his duty to impose that morality in the name of the university.

SALARY

Goodnight was in the office as Dean of Men for 29 years, from 1916 to 1945, when he was named Dean Emeritus. Handwritten personnel records track Goodnight's appointments and respective salaries over the course of his tenure at Wisconsin. Goodnight was paid $600 as an Instructor of German in 1901, a salary that increased to $1000 in 1904 and $1400 when he became an Assistant Professor. As an Associate Professor in 1913, Goodnight was earning $2250, plus an additional $500 for his role as Director of the Summer Sessions.

In 1914, Goodnight became Chair of the Committee on Student Life and received a salary of $2250. In the same year, he earned $750 for his faculty role as Associate Professor of German. In 1915, Goodnight earned $2250 as Chair of the Committee on Student Life, $750 as Associate Professor, and another $550 as Director of the Summer Sessions for a total of $3550. His appointment as Dean of Men in 1916 earned him a new salary of $3000. In that same year, he received a raise for his Summer Sessions salary of $450, bringing his summer salary to $1000. In 1916, Goodnight made the princely salary of $4000 for his 12 months of work, a significant sum in many respects.[17]

In 1926, Goodnight was earning $2000 as Director of the Summer Sessions and $4000 as Dean of Men for a total of $6000, the equivalent of $73,453 in 2010 dollars. Goodnight retained his faculty position as Associate Professor of German through 1925, according to his personnel records. After 1925, there is no further mention of the faculty appointment. Goodnight's salary as Dean declined a bit in the 1930s. However, his salary as Director, and later Dean of the Summer Sessions in 1931, increased by almost the same amount, leaving him without an overall decline. By 1938, Goodnight was still earning roughly the same amount, $6000, but the value of a salary had increased to $92,207 in 2010 dollars, due to the Great Depression of the 1930s.[18]

Salary considerations notwithstanding, Goodnight's dedication to the university was not limited to his roles as Dean of Men, Director and later Dean of the Summer Sessions or as a German professor. In 1921, he took on the added responsibility of running the campaign to raise money for construction of the Memorial Union Building on a part-time basis. He also arranged to be on a leave of absence without pay from October 1, 1928, to June 1, 1929, to be president of the Floating University, a precursor to the Semester at Sea concept. However, the Floating University never came to fruition and Goodnight stayed in Madison.[19]

In these same time periods, several university presidents came and went. Goodnight persisted despite the revolving door on the president's office, a matter of some note as presidents, like other executives, are often inclined to replace existing members of the cabinet upon their arrival. From 1903 to 1918, Charles Van Hise served the university as president and put in place both the Dean of Women and Dean of Men. Van Hise was followed by Edward Birge, who held the office from 1918 to 1925. Glenn Frank took over the presidency in 1925 and remained in office until 1937. George Sellery served as acting President until the appointment of Clarence Dykstra late in 1937. Dykstra remained in office through 1945, when he was succeeded by Edwin Broun Fred, who held the position until 1958. All told, Goodnight served as Dean under five different presidents, a considerable accomplishment in its own right.[20]

CORRESPONDENCE

As Dean of Men, Goodnight had strong opinions about conduct and comportment on campus. He saw a clear line between acceptable behavior and violations warranting disciplinary action. Letters to or about individual students from Goodnight in the early 1920s give some indication of his responses to specific incidents. These letters are typed, either by Goodnight or from his handwritten notes, and cover incidents from dropping out to fraternity membership to faculty inquiries.

In a letter dated November 9, 1920 to a parent in Pulaski, Wisconsin:[21]

> Dear Sir:
>
> On October 6, your son, James, left the university, leaving word that he was recalled by his parents. On October 15, his roommate,

Mr. M. ___, had a letter from the lad in which he said he would return in a few days. That, however, was almost a month ago and nothing has been heard from the lad since that time. Some of his belongings are still in his room at 311 S. Brooks St. and Mrs. Johnson, his landlady, is seriously in doubt as to what to do. Something like three weeks roomrent (sic) have accumulated since James went away, and she has had no word from him....

Will you be good enough to see that some action is taken at once on the matter, and that it is not allowed to drag on in this indefinite fashion....

Dean of Men

In a note to Dean Goodnight, Ernest Lane (a mathematics instructor) cited a student in his class who indicated that he would have little time for math as he was being initiated by Triangle fraternity that week. He was not allowed to shave, had to enter the fraternity through the coal-chute, and would not be allowed to bathe all week. He had also been up the night before shining shoes until 2 AM. Goodnight's response was as follows:

Dear Mr. Lane:

I want to acknowledge the assistance you have rendered in handling the case of K. E. Your note regarding his reasons for not purchasing his textbook and his statement of his attitude towards the course was of very great service to me in proving the boy's (sic) thoroughly bad situation. As you doubtless have been informed, E. has been dropped from two directions. I suspended him indefinitely for bootlegging and the executive committee of the College of Engineering has dropped him for poor scholarship.

Furthermore, your letter concerning the Triangle Fraternity and the interference of its initiation stunts with the work of its pledges formed the basis of both a conference with the president of Triangle and a circular letter to the fraternities of which I enclose a copy. I think it will do much good.

Dean of Men

Another 1920 era letter from Goodnight responded to an inquiry from the Director of Physical Education at Pomona College in Claremont, California, about activities between men in the Freshmen and Sophomore classes at the beginning of the year.

April 12, 1921

Dear Sir:

The Freshman-Sophomore Rush takes the form of a bag rush which is held on the lower campus the first or second Saturday afternoon of the fall semester. The upper classmen select a police force of 150 to 200 men who are armed with clubs and badges to preserve order and umpire the game. Fifteen to eighteen big bags stuffed with hay are set up in the middle of the lower campus. Sophomores are lined up on one side, freshmen on the other. At the firing of the gun, each side rushes forward and attempts to capture and carry back of its goal line as many sacks as possible. The play is often pretty rough. Since the freshmen outnumber the sophomores, the latter are always seeking advantages which to most of us appear like pretty poor sportsmanship. They lay tripping wires, soak the freshmen's side of the field with a hose the night before and attempt to beat the gun, etc.

We are going to try to induce the men to take on a fair contest, or a series of field day contests under the supervision of the physical education department as a substitute. I do not know how we shall succeed.

Dean of Men[22]

An exchange of letters in 1921 between Goodnight and a faculty member in Engineering indicated Goodnight's concern for the reputation of the university but also his concern for a capital campaign to raise money for the Wisconsin Union; a campaign for which he, Goodnight, had assumed much responsibility. Goodnight's letter to Professor Mead appears first, followed by Mead's reply.[23]

December 8, 1921

Professor Mead,

May I call to the attention of yourself and to the members of one of your classes an incident which produced an unfavorable impression of this University and of University students on the minds of many people. On a geology field tour this fall, one hundred students in your charge descended from the train at Ablemans, Wisconsin. Upon reaching the platform, the boys gave voice to the customary Engineering yell, "Well, well, well, Is this Ablemans? Oh, Hell!" A

Madison businessman who was on the train tells me that he never knew the thermometer to go down so fast. A group of section hands near the front end of the train threw down picks and shovels and began cursing the students in the University.... A preacher who sat opposite him [the businessman] opined that Bryan was right—it was a thoroughly godless institution. A couple of traveling men who sat ahead of him began to discourse on the uncalled for insult to the small town. My friend was of the impression that over half the people on the train went on their way with a thoroughly unfavorable impression of our students and the institution and that they would probably voice this impression and spread these sentiments wherever they went.

The incident itself is a small one but its results are certainly not unimportant. If the boys had given the Varsity Locomotive for Ablemans or had sung a stanza of "On Wisconsin," the impression would have been totally different, but to step off the train and deliberately insult the community made everyone angry who heard it. I have no doubt the boys would respond if asked about the matter, that it was thoughtlessness, and that is probably the exact truth. But why should the University students be thoughtless in a matter which so vitally affects the welfare of the institution? At the University, they are supposed to be taught to think.

In our contact with parents, with school men, and with citizens in general throughout the state, we encounter an immense amount of prejudice, misunderstanding, and hostility toward the University. We do our best to combat it and in our present campaign for a Wisconsin Memorial Union Building, we are endeavoring to mobilize the love and loyalty of our own alumni and former students throughout the state and to correct many misapprehensions which exist in the minds of the people at large. But unhappily destructive work is so much more easy and much more successful than the constructive work. I have not the slightest doubt that the Ablemans incident did more damage in one minute than we shall be able to counteract in many weeks.

I should be glad if you would bring this matter to the attention of the class concerned.

Very sincerely yours,

Dean of Men

Professor Mead replied on December 14, 1921:

Dear Dean Goodnight:

I have your letter of December 8 calling my attention to the unfortunate incident of a discourteous yell given by a class of engineering students under my charge at Ablemans this fall. I agree with you that it was a thing that should not have happened and should be prevented in the future. I am glad that you have brought it to my attention, and assure you that I shall endeavor to prevent any repetition of that sort of thing.

I would not be living up to your expectations, however, if I did not disagree in some particular with your letter. Professor Stiedtmann and Mr. Corbett were with me that day as assistants, and have a clear recollection of the incident. This was not a concerted, organized yell by one hundred engineering students as your informant implied, but a spontaneous yell by a small group judged by all of us to have been probably less than twenty. My impression, from my knowledge of Ablemans people (who for many years have sold peanuts and apples to thousands of students in geology from this institution, Illinois, Chicago, and Northwestern) is that they were not insulted, shocked or even concerned over that particular yell. They know students and understand them just as well as Madison people do. I am amused at the amount of detail regarding your crew of laborers, their action and language, which your informant was able to recall. I imagine he packs a nice little hammer himself.

I am not trying to argue for the propriety of the yell. In fact, I think the matter should be taken up with the College of Engineering "higher up" with the idea of eliminating it from the repertoire of the engineering students. I do think, however, that you have overestimated the amount of damage done by this particular incident.

Very truly,

W. J. Mead

At times, Goodnight's work was genuinely difficult. In December 1925, he wrote to the parents of a young man who had drowned in Lake Mendota, the large, beautiful lake adjacent to the Wisconsin campus.

December 20, 1925

Dear Mr. and Mrs. C. :

I write in response to your request for details of the awful accident which has cast a gloom over our community.

Coburne and Meiklejohn [another young male student] were skating on Lake Mendota [a large lake adjacent to the campus] Saturday afternoon. The lake is six miles long and four miles side to side and the basin is from forty to ninety feet in depth. The lake had recently frozen and while the ice on the bay and near shore was safe, that was not the case over the basin. The young men skated far out past the middle of the lake. They were over two miles from the town shore but the location is only vaguely known. The ice broke and both men fell in. M. managed to escape unaided and brought the news to the shore. He went out with searchers today (Sunday) but could not identify the spot at all. Six helpers have broken through the ice today in attempting the rescue work and have had to be rescued themselves. The rescue work or rather the search is being carried on by the life saving crew from the University and the Police Department. The search today has been to no avail. The holes that were in the ice have again frozen over and they have little knowledge of the place where the tragedy occurred. The search will be continued for some days, but the old heads are beginning to say that it is useless until the ice goes out in the spring. We hope, however, for the best.

I can hardly express to you my sense of the awfulness of this tragedy. It has cast a pall over the University community and we are a unit in expressing to you our deepest sympathy in this dark hour.

Goodnight's tone was both sympathetic and generous. He conveyed enough information to describe the events but with his typical economy of emotion.

Like Dean Clark at Illinois, Goodnight was himself a fraternity man. He believed in the value of the fraternity on his campus and did his best to improve them, but often to little avail. As a campus leader with responsibility for fraternities, Goodnight often heard from George Banta, publisher of *Banta's Greek Exchange*, the national newsletter and a champion of fraternity affairs. Banta's home offices were in Menasha, Wisconsin, so he was especially alert to activities in Madison. One such exchange is replicated here.

Letter dated June 14, 1923, from Banta to Goodnight

Dear Sir:

I observe in an evening Milwaukee paper the enclosed news item concerning doings at the University. I do not make mention of things of this sort in the Greek Exchange but I do like to have a slant of the real facts and I find that the sensationalism of the American newspapers does not permit them usually to give the facts in accurate form. Will you please return me the item with such comment as you care to make and much oblige.

Cordially yours,

George Banta

Goodnight replied on June 18.

My dear George,

I think the clipping which I re-enclose you herewith is rather unusually accurate. The inaccuracy lies in the headlines as is often the case. No probes or attacks have been announced by either Dean Nardin or myself, and we have steadfastly declined all along distance invitations to do so. I think the rest of the article is pretty accurate. There was a very considerable increase in drinking in the last weeks of the semester and particularly at the close of the examination period. No arrests were made so far as I know and I did not succeed in getting hold of specific instances. It was the customary sort of thing– to go over to Little Italy at two o'clock and buy a quart of moonshine, howl around until 3 to 4 A.M. until somebody gets up in disgust and comes out and puts you to bed. I hope we shall not have so much of it in the Summer Session.

Very sincerely yours,

Director[24]

RACE AND DISCRIMINATION

While Madison, Wisconsin, was not home to a racially diverse population during Goodnight's tenure, the issue of race did come up from time to time. In a particular case, Goodnight responded to a group of businessmen in Chicago who were interested in sponsoring a young man at Wisconsin. Goodnight's response was as follows:

> Letter dated September 13, 1923
>
> Mr. A.E. Geigley
> Chicago, IL
>
> Dear Dean Geigley,
>
> I am sorry to seem to discourage the entrance of a fine, young, colored boy with a good mind and ambition, at Wisconsin, but what your colleagues's son reports to you is virtually true. We have so very few colored people in the University—not over two or three men at any one time among 7,300 students. When I came to the University twenty-two years ago, I remember that a mulatto pitched for the varsity in the spring of 1901. Some years after that—1910, I think—a tall, rangy colored lad by the name of Pogue was a member of the track team and a good hurdler. These two instances are the only ones I recall in twenty-two years on the campus in which a colored boy has made an athletic team, nor have they ever participated, so far as my knowledge goes, in the general campus activities. Not that there is any rule against it, you understand, but simply that it hasn't been done. We have an occasional colored boy who graduates and I should be glad to see Mr. Bratton have his collegiate training here with us, but I could not honestly encourage him to believe that he could participate freely in student activities. He would certainly have to force recognition by sheer merit if he accomplished it.
>
> As I think over the conference athletic teams of the last ten years, the University of Iowa is the only one which I recall has played colored boys. The giant tackle Slater was one of the famous football men of the country two years ago. They have a mulatto who is a track start in the present time, but whose name I forget. I suspect however that the situation at the other Big Ten institutions is fairly comparable to what it is here.
>
> By the enclosed estimated budget of a student's expenditures, you will observe that all out-of-state students are required to pay a non-resident tuition fee of $124 at Wisconsin. The last sentence of your letter to the effect that you and your colleagues are going to loan the young man his college expenses money leads me to suggest that the sum to be loaned would be probably considerably smaller if the young man went to Illinois.

Please do not understand me to imply in anything I have said that I wouldn't like to have the young man enter here. I should be. I want to place the facts honestly before you.

Very sincerely yours,

Dean of Men

The numbers of African American students by 1938–1939 had increased significantly. A sheet of names, addresses, and class were among Goodnight's papers with the title, "Negro Students—1938–1939." The number of students on the list was 25 and only included men students. If there were female, Negro students, ostensibly such a list would have been kept by the Dean of Women.

By 1943, the issue of race reached a crescendo on campus. In a statement issued under the auspices of the Sub-Committee on Student Living Conditions and Hygiene, a response to "Report of the Student Board Housing Committee on Student Housing and Racial Discrimination" submitted to the Sub-Committee on March 22, 1943, was offered. The report, which appeared in the *Daily Cardinal* attempted to assert that there were rampant cases of discrimination against "Jewish, Negro, and Asiatic students [who] were denied rooms specifically because of race, color, or creed."

However, the faculty committee (Sub-committee on Student Living Conditions and Hygiene) found little actual discrimination based on their review of the charges. Nonetheless, it was incumbent on the Sub-committee to consider the 28 cases submitted by the Housing Committee. In turn, the faculty found that most of the cases involved young female students and further, that in most cases, it was the students' disruptive behavior, not discrimination, which caused the removal from a housing unit. Nonetheless, the faculty committee report does acknowledge that there is little, if any discrimination, against Jews but that "some houses draw the color line."[25]

Goodnight's personal views of the issue of race are limited to the letter to Dean Geigley in 1923 although it is clear that he was involved in keeping track of Negro students in the 1930s and certainly involved in the concerns over housing in 1943. Despite what appears to be racist language in his letter to Dean Geigley and the use of the terms, "colored" and "mulatto" in the text of his response, such terms were not considered intemperate in 1920. What is clear is Dean Goodnight's even-handed and sympathetic concern for the well-being of the young man in question, Mr. Bratton. Goodnight clearly states that he will likely be ostracized and excluded from many activities at Wisconsin because of his race. He further addresses the issue of finances and urges the prospective student's benefactors to save money by attending an in-state university in Illinois.

DISCIPLINE AND THE DEAN

Despite many of the actions, letters, speeches, and moralizing that can be attributed to Goodnight, in fact, the university had drawn a broad line between both the Dean of Men and the Dean of Women when it came to direct responsibility for student discipline. No student was sent directly to the respective dean for disciplinary action. Instead, both Goodnight and Louise Nardin, who had replaced Lois Mathews as Dean of Women, were buffered by the University Committee on Student Welfare, a committee composed of faculty and administrators who actually heard disciplinary infractions and meted out the punishments, suspensions, or expulsions as necessary. It was determined in the early 1920s that no dean who hoped to work with students could also be the primary person responsible for discipline along with advising, counseling, and similar tasks.

Goodnight served on the conduct committee and had even been chair before his appointment as Dean of Men in 1918, but he made sure he did not have direct responsibility for student discipline as Dean Clark of Illinois did. This pattern of responsibility became problematic for many deans of men, who often saw conflicts between their role as disciplinarian and advisor. It was difficult to do both well; a conundrum that would persist for many decades.

In fact, this arrangement was called into question in 1930 when the *Daily Cardinal*, the campus newspaper, published three separate editorials on the matter of student conduct and the role of the Deans of Men and Women. The *Daily Cardinal* editorialized that the university should remove the role of disciplinarians from both the Deans of Men and Women's responsibilities. Instead, the paper argued that students should be subject to the rules of law imposed on "ordinary citizens" by the local police and not the "police powers" ascribed to the Deans.[26]

The paper also argued in a third editorial that the university should employ a psychiatrist for the assistance of students who presented any issues of mental hygiene. The paper was concerned that the only recourse available was through disciplinary measures, which the *Cardinal* editorial viewed as inappropriate. Although the *Daily*

Cardinal believed so strongly in its position that it devoted a separate issue to reprinting all three editorials on the front page, the paper did not succeed in dislodging Goodnight or Nardin from their roles as disciplinarians or arbiters of student conduct.

THE GRAY BOOK

Goodnight saw his role as dean as arbiter of moral values, especially for young men, but often, for all students at Wisconsin. Like Clark at Illinois, Goodnight used the student press as an avenue for his teachings on behavior, conduct, and the moral strictures of college life. During orientation programs for new students, copies of the university Gray Book, a guide to student life, were distributed to all students. The Gray Book was a colloquial title, referring to the grey-colored paper that bound the book. Officially, the Gray Book was titled, *Regulations for the Guidance of Undergraduate Students.* Under the title on the title page were printed the words, "Students are held responsible for knowledge of the contents of this booklet." The 1921–1922 Gray Book covered topics as diverse as attendance and "marks" (grades) to physical education (a university requirement for all freshmen and sophomores) to regulations governing extracurricular activities.

The publication of the Gray Book was an opportunity for the press, both on campus and off, to peek behind the curtain at university life in the 1920s. The campus paper, the *Daily Cardinal*, duly cited Dean Goodnight in its story headlined, "Gray Book Explains Life at University to Incoming Freshmen" (*Daily Cardinal*, August 5, 1926). "In 2000 homes, the 1926 crop of university freshmen is this week being introduced to the environment which its members will enter next month. The medium of introduction is the Gray book, published to give advance information to the freshmen concerning campus life."

Other newspapers, including the Madison local paper, took note of Goodnight's battle with local "road houses," which were replacing the "old time saloons" in the city. The paper cited Goodnight charging that the road houses were "a paradise for hippers, bootleggers, and people who are out for a blowout." In concert with the Dean of Men in the condemnation of illegal and improper alcohol use, Dean of Women Louise Nardin wrote in the 1927 edition of *If I Were A Freshman Again*, a booklet given to all freshmen women, that the code for women at the university included, "4. She refuses to associate with men who have been drinking." (p. 5).

SUGGESTIONS FOR PARENTS

Goodnight also took pen in hand to draft "Suggestions to Parents" for the parents of new students. Among his suggestions were discussions on money, cars, and athletics, especially football. Some excerpts from Goodnight's 1926 "Suggestions to Parents" offer the flavor of his advice.

> Financial status has much to do with the success or failure of a student in college. The ideal is, enough money to support him comfortably and to allow full time for his work and his recreation. Too little money compels him to expend so much time and energy in work for self support that his college work must suffer.
>
> ...But too much money is worse than not enough! (sic) A surplus of money beyond actual needs is a standing temptation to excessive indulgences in dances, joy-riding, out-of-town excursions, convivial nights, gambling, and the whole array of amusements which appeal so strongly to youth and which, if cultivated to excess, so effectively distract a student's energies from the true purpose of college life.
>
> With rare exceptions—e.g. students who have a physical disability, or who live at home but outside the city of Madison—no student needs a car at the University! (sic) The reasons are many and obvious: it is a source of real danger to life and limb, it costs more in time than the wealthy student and more in money than the poor student can afford, it is a constant distraction from study, even if it be put to no worse than joy riding, and the temptation to put it [the car] to worse uses is ever present.
>
> Student migrations to universities in other states to attend football games is a practice which should not be encouraged. There are four or five home games each fall, which provide excitement enough. The migrations seriously disrupt the college work. There is always drinking and bad conduct when great numbers of students go on such an excursion. The cost of such trips is not inconsiderable.[27]

CAMPUS ORDER

As a campus arbiter of student conduct and behavior, Goodnight was often called upon to confront student behavior that ran counter to university regulations. Everything from drinking, gambling, illegal cars, academic misconduct, and fraternity hazing was put on Goodnight's plate. One of the most famous stories attributed to Goodnight was that he once waited all night on the porch of a Madison boarding house to confront a man and woman, both students apparently, who had spent the night together. Although the alleged incident took place in the 1930s, Goodnight found himself correcting the story even as late as 1960, long after his retirement.

The incident was so notorious a legend that Goodnight wrote a final correction to the *Wisconsin State Journal* on the occasion of his eighty-fifth birthday. In his own words, written in a letter to the editor long after his retirement in 1945, Goodnight tried one last time to put the issue to rest.

To the Editor, *The State Journal*—

> The most memorable birthday I have had has been marred for me by the publication for the umpteenth time of the old libel about my having "sat before the door through the night until the next morning," etc. in the *Wisconsin State Journal.*
>
> When it first appeared in 1931, I wrote to the *Wisconsin State Journal* a faithful and true account of the whole affair but it availed not, and in every article since then, The *Wisconsin State Journal* has printed about it, the old lie is repeated.
>
> The dean of women alerted me to "a case of student misconduct" in the Irving apartments shortly after 8 o'clock one morning, just after I had arrived at my office. I went to the apartment indicated and you report with fair accuracy the dialogue.
>
> The "rocking chair vigil" was just as you state it, except for the fact that it lasted about 20 minutes and I was back in my office before 9 o'clock.
>
> I may not always have acted wisely as dean of men but I assure you that, even in my "balmiest days," I would never have done such an asinine thing as to sit all night before the door of an adulteress couple.
>
> I wonder if there is any means available to me by which I could induce *State Journal* writers to cease that old lie that has plagued me for 30 years? I now regret that did not bring suit for libel against the papers and the AP and the UP that broadcast it. I am sure I would have won sizeable damages.
>
> But your statement that I customarily shrugged off criticism is quite true and I did so in that case.
>
> (Signed) Scott H. Goodnight, 1649 Aloma Ave., Winter Park, Fla.[28]

The salaciousness of the incident, especially in 1931, was so intriguing, as was Goodnight's alleged response, that few, if any, newspaper reporters or alumni, wanted the legend to die. Clearly, the incident spoke to the vast span of control over student lives assumed by both deans of women and deans of men in the early days of coeducation. What makes the story so much fun is the idea that Goodnight would wait out the miscreants by sitting in a rocking chair on the porch of the boarding house. Clearly, Goodnight was not amused by the story. But despite his best efforts, the story survived intact in the minds of students, alumni, and the *Wisconsin State Journal.*

Over the course of his career at Wisconsin, Goodnight adapted readily to a wide range of changes both culturally and administratively. In his time, he survived five university presidents, a daunting task in and of itself. Goodnight was deemed to be successful in his office as he retired of his own volition in 1945, serving the university for a period just shy of 50 years, beginning in 1901 as an instructor of German and followed by his appointment as Assistant Director of Summer Sessions in 1911 and then Dean of Men in 1916.

During that time, Goodnight saw the end of World War I, increases in student enrollment, the Depression of the 1930s, and the turmoil of World War II. He was instrumental in the development of a professional association for the deans of men, the National Association of Deans of Men, and served the association as secretary for three terms and as president for one. He represented the University of Wisconsin ably in state, regional, and national circles. Goodnight and Louise Nardin had a good working relationship, in part because they both assumed their positions in the same year, 1916. A good bit of correspondence was exchanged between the two; in fact, the notorious boarding room "rocking chair" incident in 1931 was a direct result of Goodnight's response to the Dean of Women who reported the incident to him.

Despite the rancor that sprang up, on individual campuses, between the professional associations of deans of women and deans of men in the 1930s, the two offices and the incumbent deans at Wisconsin often worked together harmoniously. Goodnight demonstrates little direct contact with female students as Dean of Men although he chaired the Committee on Student Life and Interests, a committee that addressed student concerns and issues of all kinds. Dean of Women, Louise Nardin is listed as Assistant Chairman, in correspondence in the 1920s, further evidence of a reasoned and public working relationship between the two deans.[29]

Over the years, Goodnight did appear to mellow a bit as the job got easier or more likely, he got better at it. In a radio address delivered in 1922. Goodnight reflected on the challenges of campus life in the post-war period from 1917 to 1920, when in his words,

> The question of youth, and especially college youth, has come into the foreground of public attention. Every college officer, whether

> president or dean, is frequently asked to go on record with regard to his opinion of modern youth. Are the students of today so much more wicked and irresponsible and laze [sic] and dumb than the students of two or three decades ago? ... To this question, I have no hesitancy in replying. Be it understood, however, that I am speaking of University of Wisconsin students only.... In my opinion, the students of today are distinctly superior to those of two decades ago, in both scholastic achievement and conduct.[30]

By the 1940s, Goodnight had fixed his position even more firmly in the positive. In a radio interview, he responded to a question about comparing students in the 1940s to those of the previous generation with the following observations:

> The young people come to us much better trained and prepared than in the old days. There is much less verdancy and crudeness in the incoming freshmen class. As they go through college, they meet stiffer requirements and they measure up to much higher standards of scholastic work. They live in a vastly more complex world and they are completely at home in it... These young people are, for the most part, alert, frank, fearless, and well-poised. It is a rare privilege and a stimulating experience to live among them and to work with them.[31]

Despite his age, the strong sense of righteousness and indignity were never far below the surface. In an exchange of letters with John Edgar Hoover at the Federal Bureau of Investigation in the 1940s, Goodnight's sense of moral outrage came though when he contemplated the threat of "communists" infiltrating his university.

> Letter dated May 1, 1940
>
> My dear Director Hoover:
>
> Permit me to thank you for a copy of your splendid address "The Test of Citizenship" delivered before the D.A.R. I sympathize with it 100 per cent. Tolerance is a fine thing in national, religious and political matters but in my belief it has its limitations too and with you I draw the line on the rats (sic) that are undermining and seeking to overthrow our whole form of government and our whole civilization. More power to you and be assured of my backing whenever I can say a word for you.
>
> Very sincerely yours,
>
> Dean of Men

A subsequent newspaper article reported "there are more than 11,000 students at the University of Wisconsin and you could put all the Communists on the campus in one end of a box car for convenient shipping back to New York," Scott H. Goodnight, dean of men of the State University, said in a recent radio broadcast from the university campus." Goodnight was further quoted directly as saying, "My guess is that there aren't more than 30 or 40 Communists in the whole student body. There is a wholesome, normal spirit of youthful liberalism on this campus but it is on the whole tempered and sane."[32]

Later in the same article, when asked to characterize the changes in students from his first years on campus, beginning in 1901, Goodnight replied, "Boys today, as well as girls, are more sophisticated than their predecessors of 40 years ago. The know more of both the good and the bad things of life; they are better read and better informed in foreign affairs, statecraft, economics, science, and even in literature and art, but they are also more bohemian. . . ." Goodnight also informed the reporter, "These young people are, for the most part, alert, frank, fearless, and well poised. It is a rare privilege and a stimulating experience to live among them and work with them."[33]

Nearing the end of his career at Wisconsin, Goodnight was able to look back on his work with a profound sense of accomplishment and gratification over a 40-plus year career as an educator and administrator. He had stopped teaching by 1927 and gave up his second career as Dean of Summer Sessions in 193 ? He saw his duties as Dean of Men expand as the enrollment climbed in the 1930s and 1940s, especially as World War II ended. But Scott Goodnight timed his exit well and retired to Winter Park, Florida, in 1945, escaping from the rapid expansion of American colleges and universities that blossomed in the late 1940s and early 1950s.

Scott Goodnight's career served as a barometer for the changes in the work of deans of men. As he was present at the very first gathering of deans at Wisconsin in 1919 and was witness to many changes in the profession by the time his career as a dean ended in 1945, it is possible to trace the evolution of the profession of dean of men through Goodnight's activities.

In 1919, when Goodnight had invited the other deans to his campus for a discussion, the deans of men were struggling to determine what their roles were. To a large degree, they followed the work of their mentor, Thomas Arkle Clark at Illinois, in charting the direction for their own campuses. Scott Goodnight took on his work as the Dean of Men as a part-time position in addition to his role as a faculty member in the German department and his summer employment as Director of the Summer Sessions.

As Dean of Men, he gave radio addresses, kept numerous records on class attendance, granted permissions for social activities, and kept a lid on the social antics of the fraternities and other student activities, and maintained the inspections of off-campus housing, most of the work accomplished with only his office staff of two or three women and Goodnight as dean. But by the 1930s, Goodnight was petitioning the university presidents for additional staff and money. The role and responsibilities of the Dean of Men had expanded along with the enrollment at the university. The scope of responsibility had expanded as well.

For example, as Goodnight notes in his report in 1931, several hundred files for all initiates into the 52 fraternities were kept annually. In addition, "all social functions, banquets, dances, mixers, and the like must be registered in advance and approved. Those at which both men and women are in attendance, must be properly chaperoned and the names of the chaperones registered in advance.... During the academic year, 1929–1930, there were 520 dancing parties (exclusive of Prom and Military Ball parties or commercial dances each Friday and Saturday night)."[34]

The office of dean of men at Wisconsin also inspected and kept lists of all lodgings available for male students as well as "flats, apartments, and houses for families. This involves the inspection of from 1200 to 1400 lodging houses each year, of which an annual report is submitted. A file card is kept for each house, showing capacity, prices, grade, names of lodgers, etc."[35] Much of this work was done by Mrs. Blance Stemm, formerly a "trained nurse, she knows well the importance of cleanliness; she is tactful and intelligent and can argue convincingly when necessary. Fortified with her own inspection reports, she can usually tell a complaining student more about his lodging house than he himself knows."[36] Goodnight's office also employed an accountant who served as the student financial adviser and oversaw the accounts of all student organizations on campus. Managers and treasurers of the organizations were placed under bond and had to submit statements of their financial transactions annually. Student loans were distributed through the Dean of Men's Office as well.[37]

As the "personnel" movement advanced, deans of men and women began to conduct more and more interviews with individual students. New, incoming freshmen students were called in for interviews, during which time a personnel card logging basic information on each student was prepared, to be maintained during the student's collegiate career. Additional notations could be made as needed and, more importantly, a record of each student's progress could be reviewed at key points during their time in the college. Goodnight himself conducted the individual interviews with the new male students at Wisconsin. In addition, he held individual meetings with male students on a variety of issues, ranging from student conduct to student loans to general advisement and information. In 1930, the number of interviews reached 1300 during the first [Fall] semester or an average of about 19 per day. Goodnight's report notes that "this does not include perhaps three times that number of persons who came to the office during that time and were served by the young women in the outer office, but only those who had personal appointments with the Dean." In addition, the dean's report notes that he had attended 87 committee meetings in the first five months of 1930.[38]

By his own admission, Goodnight did not complete a comprehensive annual report every year on the Office of Dean of Men as such reports were not consistently requested. However, in several years in the 1930s and again at the very end of his tenure as Dean in 1945, Goodnight and his staff put together a comprehensive overview on the work of their office.

In his 1938 report, Goodnight noted,

> When, in 1916, the office of dean of men was created, no directions or advice were given, and the work was conceived merely as a further projection of that of the Committee on Student Life and Interests.
>
> In the years that have since intervened, various student welfare agencies have been established on the campus, usually independently of such other or existing agencies. The student employment services; the entire development of the Union as a separate and virtually independent organization; a bureau of guidance and records, quite unattached; the information offices in Bascom [administrative hall]; the dormitory committee, quite independent of others; these are all examples. There is no implication intended that these agencies do not cooperate: they do, very freely and helpfully, as a rule, when the opportunity is presented. The difficulty lies in a total lack of coordination (among the student services organizations—added in pencil).
>
> It is not only at Wisconsin that this situation prevails. All over the country, institutions are becoming aware of the desirability of organizing student personnel work on a more unified basis. The deans of men in their annual meetings for the last three years have given much attention to the "student personnel point of view", as it has been termed, and there are movements now afoot to promote the work of coordinating and social and educational direction, to the end that the

> individual student may receive the development that most fits him for the most effective participation in society.[39]

In the report, Goodnight enumerated the various activities of the office over the previous year, including the Grey Book, freshmen convocations, and other activities. He made a strong case for the expansion of his office as well, citing the lack of growth in the office and in staffing despite significant growth in enrollment over the same time period. What is interesting to read in the 1938 Report is that although Dean Goodnight recognizes the advent of the "student personnel movement," he never returns to the discussion beyond the first few pages of his report. He acknowledges the change that is occurring and that will some day become standardized at Wisconsin but in fact, he has not quite given himself over to the idea, much less embraced it as his own philosophy of practice. In his defense, at the age of 63, Goodnight could accept the new concept of the "student personnel point of view" but was unlikely to embrace it as practice.

Despite the administrative changes in the office of the dean, the problems remained much the same, no matter the year. At the conclusion of his 1938 report, Goodnight commented at length on the issue he still found most troubling—The Drink Problem.

> The insoluble student problem of the ages is the drink problem. We are no further along with it than were the medieval universities of Europe. Beyond the slow and often ineffective process of attempted education, there is little that can be done about it. In the first place, drink was ever lawless; regulations, like the eighteenth amendment [which created Prohibition] are useless, because they are consistently violated. Further, the people of Wisconsin voted liquor back by an over-whelming plurality; the legislature was liberality itself in framing the new liquor laws: neither legislature nor city council would consent to restore the half-mile dry zone about the campus; there are taverns at every turn; and finally, to complete the picture, the Regents voted beer into the Union![40]

Goodnight was not railing against liquor in only moral terms, although his ministerial heritage is clearly peeking through. He rails against the common foe he has battled for his entire career as a Dean of Men, alcohol. He now, as he notes in the annual report, has to contend with the argument from every fraternity party that is brought to his attention that the "Regents are selling beer in the Union" so why can't we? Goodnight argued, with good evidence, that most of the behavior problems he encountered over time were often the result of the "drink problem." (For those familiar with the University of Wisconsin Memorial Union, the problem still persists as the Rathskeller, a large pub on the first floor, is still operational year round.)

By 1945, at age 70, Scott Goodnight was ready to say farewell to the office he had helped create in 1914 and then occupied from 1916 on. He wrote a final, extensive report, outlining the many duties and responsibilities that had accrued to the Dean of Men over time, in hopes that his successors might have a path to follow, something he had to create on his own. In his own, modest way, the title of this "final" report speaks to his own philosophy as dean. The report is titled, Organization, Scope, and Practice of Faculty Control of Student Extra-Curricular Activities and sub-titled, The Committee on Student Life and Interests, [and] Its Subcommittees, The Dean of Men and the Secretariat of the Office of Dean of Men. The quintessential point, from Goodnight's point of view, was that these responsibilities were not those of a single man, but rather the responsibility of a committee of faculty of the university who, in turn, authorized the Dean to act on their behalf.

Although Goodnight had not been a teaching faculty member since early in his career, he was still adamant about placing the role of the dean of men and dean of women within the proper context. In his words, the control of extracurricular activities was under the purview of the Wisconsin faculty. Goodnight was determined to make this point even as he was about to depart the university. In part, his concern was for the continuation of the work he had been engaged in for almost 30 years.

Goodnight was also the consummate bureaucrat. The word, bureaucrat, is used in the most positive way. Goodnight saw a clear precedent and an institutional history to be observed in the organization of the Committee on Student Life and in the development of the Office of Dean of Men. The scope of authority for these duties was embedded in the structure of faculty governance. Scott Goodnight was determined to ensure that those conditions continued to be observed long after he had left Madison.

At 70, Scott Goodnight was finally ready to leave Madison, Wisconsin and the University of Wisconsin, his home and employer for almost 45 years. His departure was not without many celebrations and a campus wide appreciation for his many years of service. He gave radio addresses and interviews, and was celebrated long after he left. He had certainly earned his retirement, a retirement he was able to enjoy for the next 27 years. Goodnight and his wife, Gertrude, relocated to Winter Park, Florida, where they enjoyed a life of leisure and rest. On several major birthdays, correspondence reached the campus and celebrations of Goodnight's birthdays, in particular, his eighty-fifth and ninetieth were announced in press releases and newspaper accounts through newspapers in Madison and Milwaukee. On his 90th birthday, Goodnight apologetically had to resort to a mimeographed letter to respond to the many well wishers who had sent him cards and letters. But he made it clear that he had no desire to return to Madison. "I have no yen to return to Madison, despite the number of good friends I have there. It makes me shudder to think of 150,000 souls (19,000 when I first knew it in 1901) and the University, then with 3,000 students, now with—?"[41]

Goodnight lived to be 97. In 1965, plagued by failing eyesight, the widowed Goodnight moved from his home to a newly constructed retirement home operated by the Presbyterian Church in Winter Park, his home until his death in 1972.[42] In an interview in 1965, Goodnight acknowledged that he had not been in Madison since 1959 and at age 90, he "doesn't want to see Madison again, period."[43] Goodnight told his interviewer that the university had grown too large for his tastes.

Unlike his fellow dean of men, Thomas Arkle Clark, Scott Goodnight had not sought out the limelight, either on campus nor as a nationally known speaker and author. For most of his career, Goodnight confined his ambitions to the campus at Madison. A few forays into interesting and perhaps more lucrative ventures can be noted. At one point in the mid-1930s, Goodnight had applied for a leave of absence to serve on a floating ship at sea; he referred to it as his "Ivory soap university," but the deal apparently fell through and Goodnight stayed on the job in Madison.[44]

Scott Goodnight was the quintessential dean of men from the first half of the twentieth century. He smoked a pipe, graduated from the same university where he worked for almost 50 years, had a gruff exterior but a deeply compassionate nature, especially for the young men whose lives intersected with his own for many years. He railed against change and berated slovenliness, cheating, excessive use of alcohol, and aberrant behavior.

Clearly, he touched many lives and lived on in the memories of students, faculty, and staff. Despite the fact that he had retired in 1945, his eighty-fifth birthday celebration was chronicled in the *Wisconsin State Journal* with a three-column story and a large photo of the Dean. His ninetieth birthday received almost as much attention five years later. As the *State Journal* reported, Goodnight himself characterized his years as dean as, "the only satisfaction I got from bearing down on a boy is when they come back, sometimes years later, and told me that they have come to realize that the medicine was good for them."[45]

4

The Academics: Early Deans in the Liberal Arts Colleges

LeBaron Russell Briggs

LeBaron Russell Briggs was the man at Harvard College called on to take on tasks no one else wanted to do or, in some cases, no one else thought of doing in the first place. Briggs, often referred to as "Dean Briggs" by the students, became much more than his preferred professional occupation as professor of rhetoric over his long career. In time, Briggs would actually become Dean for Students. Briggs was so instrumental to Charles Eliot's administration of Harvard College that he also served as the President of the Harvard Annex, better known as Radcliffe College, for several years.

But Briggs' acclaim among Harvard students relied on unofficial activities that did not warrant an official title or a sign on his door. His true role at Harvard was to be father, trusted confidant, beloved uncle, confessor, minister, priest, and big brother to years of Harvard undergraduates. Long lines of freshmen would form outside his office door, waiting to get his wise counsel on courses to take, clubs to join, or just to have a friendly chat about missing home and family or a lack of available funds.[1] Briggs' importance to the men of Harvard College was recognized at his last commencement in 1925, just prior to his retirement, when President A. Lawrence Lowell announced that an endowment fund of $63,490 ($775, 000 in 2008 dollars) had been established by Harvard alumni for Dean and Mrs. Briggs, "out of the depth of their respect, loyalty, gratitude, and affection for Dean Briggs."[2]

LeBaron Russell Briggs was born in 1855 to George Ware Briggs and his second wife, Lycia Jane Russell Briggs. George Ware Briggs was a minister and intentionally moved his family from Salem to Cambridge so that his sons might attend his alma mater, Harvard College. LeBaron Russell graduated from Harvard, as his father had hoped, in 1875, fourth in his class. As a student, his mentors were Professor George Lane and George Herbert Palmer, who taught Latin and Greek, respectively.[3] After his graduation, Briggs travelled overseas and eventually returned to Cambridge, where he began to study for his Ph.D. in Greek.

In the fall of 1878, Briggs received his first overtures from Harvard President, Charles Eliot. Eliot offered Briggs a position as a lecturer in Greek for one year. Briggs accepted. At the end of the first year, his contract was extended for two more years.

Although he initially saw himself as a Greek scholar, Briggs soon became very interested in English and the related field of rhetoric. Like Eliot, Briggs saw the need for undergraduates at Harvard, especially the freshmen, to have solid grounding in composition and rhetoric. He began to move away from Greek and to pursue a masters degree in English. A despondent Briggs struggled to complete his doctoral program after his mother passed away in 1881. Upon completing his degree in 1882, he travelled to Oxford for further study, but found little that suited him, so travelled across Europe for much of the year with his father.

After returning to Harvard in 1883, Briggs was assigned to help Professor Adams Sherman Hill develop a program in English. Both Eliot and Hill believed that a solid grounding in English and good writing would serve the Harvard men well, a conclusion that Briggs enthusiastically supported. Eliot proposed the requirement for an English composition course should be met in the sophomore year, but it soon became clear to Briggs and others that it was the freshmen who would benefit the most from such a course. They convinced Eliot to move the requirement and soon, Briggs and his peers were teaching Harvard freshmen the basics of English composition.[4]

It was this early affiliation with new students that cemented Briggs' place in Harvard history. Briggs had a natural affinity for working with young men, especially the freshmen who were often alone and socially adrift on the campus. Briggs freely offered advice on classes, roommate issues, and other problems that confronted the young men. At the same time, Briggs expanded the freshman English courses and exceeded the expectations Eliot had laid out a few years earlier for a solid grounding in reading and writing skills.

Briggs' success with the English program as well as his natural charm with undergraduate students led Eliot to appoint him to be Dean of the College in 1891. When he published his book, *College Administration*, in 1908, Eliot referred to his two appointments of administrators as "a dean for the students and one for the faculty."[5] So Briggs was, in fact, the dean for students. Because there were only male students at Harvard at the time, he could also be called the dean of men. Regardless of his official title, Briggs' appointment as dean preceded the appointment of Thomas Clark as Dean of Men at the University of Illinois by several years.

At the age of 35, Briggs was a clear success. He was well established at his alma mater, Harvard College. He expanded the freshman composition program and set the tone for the teaching of English on other campuses. He was made Dean of the College in recognition of his success with students. He was well regarded by his employer, Charles W. Eliot, as well as by other faculty and administrators. Briggs brought an important student perspective to the Administrative Council meetings Eliot held and, through his efforts, influenced policy and practice on campus.

But the varied tasks of teaching and advising many students in addition to his work as Dean of the College took their toll on Briggs. His late nights and long hours exhausted him. To provide some relief, in 1902, Eliot re-assigned Briggs to be Dean of the Faculty of Arts and Sciences. The new position may have limited some of the student demands. But in 1903, when Eliot needed a new head for the Harvard Annex, later re-named Radcliffe College, he asked Briggs to take the position on a "part-time" basis. Briggs served, "part-time," for the next 10 years.

LeBaron Russell Briggs served under two powerful presidents, Charles W. Eliot and A. Lawrence Lowell, over a period of some 40 years. A man of many talents, Briggs made a significant mark on higher education in large part due to his attentiveness to the needs and support of undergraduate students. He became the role model for those who followed him, the quintessential adviser who not only could empathize with students but respond to them as a person of knowledge and wisdom, setting them on a path toward success with a new sense of self-confidence and ambition. In his own words, Briggs declared,

> Almost equally important with an understanding between parent and son is an understanding between every student and at least one college officer. There must be some one on the spot to whom the student may talk freely and fully about such perplexities as beset every young man in a new life away from home. Even a college bred father is college bred in another generation and cannot know those local and temporal characteristics of a college on the mastery of which depends so large a measure of the student's happiness"[6]

Briggs wrote extensively. He authored numerous books related to his academic field, English and was a pioneer in what has become known as the freshman composition course. Briggs also wrote and lectured extensively about his beliefs about college students. Many of his ideas are anecdotal and drawn from his extensive experiences with young men, and to only a slightly lesser degree, young women, given his years at Harvard as well as his time at Radcliffe. Briggs echoes a familiar, re-occurring theme in urging parents, college professors, and society in general to maintain high expectations for students but, at the same time, to recognize the many challenges inherent in achieving that success. He spoke of character, moral values, purpose, and commitment frequently. Indeed, the titles of his books were clear indicators of his beliefs; *School, College, and Character,* and *Routine and Ideals.*

Briggs was a champion for the young men he nurtured at Harvard. At the same time, his Radcliffe years gave him insights into the lives of young women that many Harvard professors and administrators lacked.[7] As the father of three daughters, Briggs no doubt had considerable personal experience with young women, but Radcliffe instilled a sense of immediacy and purpose for his support of educated women. In several speeches, Briggs acknowledged the disparity between male and female college graduates. Although far from an advocate for feminism, Briggs was more than a little sympathetic to the limitations society in the early twentieth century placed on a bright and capable young woman with a degree from Radcliffe, Wellesley, or similar schools. He encouraged young women much as he did young men to make the most of the opportunities that an Ivy League education afforded them and to be faithful to their own high standards.[8]

Briggs' professional model as one of the first college officers appointed to oversee the lives of undergraduate students is both instructive and inspirational. He managed it with grace, good humor, and a steadfast belief in young men and women. Many other faculty who were appointed to provide supervision for an undergraduate student population attempted to emulate Dean Briggs, but that was a difficult task. Briggs was not only a unique individual but he was in very rarified air. He was hand picked by Charles Eliot to serve first as a faculty member, later as an administrator, at one of the most elite, liberal arts colleges in the United States, and possibly the world. In his 40 years as President of Harvard, Eliot was not known to suffer fools kindly. Had Briggs been too far afield from Eliot's beliefs, Briggs would have quickly been dismissed as Dean.

Briggs' ability to provide a much needed and essential role as an intermediary between the Harvard faculty and administration and the undergraduate students is still impressive today. Perhaps it was his early childhood years as the son of a minister or his own experiences as a Harvard undergraduate that drew out the empathetic character traits that suited Briggs so well for such a daunting job. Clearly, as Dean of the College, he could be firm and unflinching when necessary. Briggs dismissed any number of students for academic as well as social infractions. But he was also capable of sensitivity, compassion, and concern that endeared him to scores of Harvard men. It was Briggs' compassion over many years that caused so many of his students to contribute from their own pockets to his welfare when given the chance.

ALEXANDER MEIKLEJOHN

Like Briggs, Alexander Meiklejohn was first employed by his alma mater. Meiklejohn was a young and proud alumni of Brown University. When he graduated in 1893, Meiklejohn debated between becoming a professional athlete or an academic. A gifted athlete with a stellar intellect, it was not an easy choice. But finally, after much deliberation, he chose the "life of the mind" and completed his masters degree in philosophy at Brown and then completed his Ph.D. in philosophy at Cornell.[9] Meiklejohn returned to Brown in the same year he graduated from Cornell (1897) and was hired as an assistant professor in philosophy. He was promoted to associate professor in 1899, and professor of logic and metaphysics in 1903.[10]

In 1901, at the age of 29, Meiklejohn was made Dean of the College, becoming only the second man to hold that position. As dean, he was responsible for oversight of discipline, athletics, and social life. Like Briggs, Meiklejohn was a well-respected faculty member prior to his position as an administrator. His appointment as Dean of the College gave him oversight responsibilities for the undergraduate population which, as with Briggs, was male only.[11]

Meiklejohn's age was an asset in his dealings with the undergraduates at Brown. He himself was not so far removed from the spirit and enthusiasm of undergraduate life, and he quickly became a favorite of the students. He had already established himself amongst the undergraduates as one of their favorite teachers, primarily through his class on logic. He approached philosophy and teaching much as he had athletics, as a competition that demanded his best efforts. And as in athletics, he often won, in this case, winning the loyalty and admiration of his students.

In the classroom and as Dean, Meiklejohn pushed his ideas and beliefs about democracy and education on students in the classroom and on campus in general. A liberal education, by Meiklejohn standards, carried with it great responsibility and obligation. He challenged the young men of Brown to shoulder more responsibility for grades, academic pursuits, and intellectual aspirations. As did many college men of his day, Meiklejohn believed in the social and even intellectual benefits of the college fraternity. But he chided the fraternities at Brown for not using their collective efforts to maintain the best grades on campus rather than engaging in social activities and antics. Meiklejohn believed Brown should represent two ideals: fair play and think.[12] By this statement, he intended that Brown students should be concerned about the democratic principles of fairness to all. By think, he meant that the young men needed to constantly consider the world they lived in at Brown but even more so, the world beyond: What was right, what was ethical, what was fair to oneself and to others.

Over his 11 years at Brown, Meiklejohn created a larger than life persona which eventually led to his appointment, at the age of 35, as President of Amherst College in 1912. As a professor, Meiklejohn, like Briggs, captured the spirit of the undergraduate soul in the classroom. He challenged students and made them examine themselves and the world in which they lived according to the principles and challenges of philosophy. The intellectual rigor in his classes was high but it was meaningful and relevant. As Dean of Men, Meiklejohn played out his ideas and beliefs in a real world setting outside the academic cocoon of the classroom. As his biographer, Adam Nelson, describes Meiklejohn's approach, he was able to apply "democratic principles" through his work as Dean.[13] One of the most compelling areas was in the awarding of financial aid. Meiklejohn saw intellectual curiosity as the primary criteria for admission to Brown, not family standing or reputation or even the new standardized exams encouraged by the NEA's Committee of Ten to evaluate college applicants. If the award of financial aid to a deserving young man might bring working class students (like himself) to Brown, Meiklejohn was all too happy to reward intellectual curiosity over social standing.

As Dean, one of the great challenges Meiklejohn faced was an issue dear to his heart, college athletics. As an outstanding collegiate athlete himself (in several sports), Meiklejohn was deeply committed to the values and benefits of collegiate sport. To Meiklejohn, athletics were another example of democracy and fair play. Participation and the opportunity to represent one's alma mater on the playing field against worthy opponents was almost chivalrous. Winning or losing was far less of an issue than the physical challenges and the joy of playing hard. But in the early twentieth century, college athletics was facing many challenges, not the least of which was the issue of a creeping professionalism and the incursion of commercialism on the campus.[14]

The major issue to face Meiklejohn at Brown was the matter of "summer ball."[15] In short, summer ball was the exclusion of amateurism in the summers when college athletes were not enrolled in college. As "independent agents" no longer tied to their colleges in the summers when schools were closed, young men could play for pay as semi-professionals and then, upon re-enrollment in the fall, become amateurs once again on their college team.

The issue was critical to Meiklejohn as it spoke to all of his key issues—fairness, democracy, and more. However, he had turned the responsibilities for the governance of intercollegiate athletics over to the Brown student government in 1906. Brown had abided by the amateur rules, but now the student government reversed that decision and voted to allow professionals (e.g. summer ball players and others) to participate on Brown athletic teams. The decision was crushing to Meiklejohn, but he had no choice but to let it stand. In a democratic process, the students had made their decision and he, too, was bound by it.

Nonetheless, Meiklejohn, at Brown, and later at Amherst and the University of Wisconsin, rose again and again to press his commitment to democratic ideals and the concepts of fairness, fair play, and liberal education. His ideals would serve him well but he was never without controversy and dissenters. Though the students at Brown (and later at Amherst) loved Meiklejohn for his "square dealings" in his interactions with them and his concern for their welfare, Meiklejohn often challenged his peers and superiors with his lofty ideals and his zeal. A brilliant professor and a great Dean of Men, Meiklejohn left Brown behind in his quest to prove himself on a larger stage.

WILLIAM ALDERMAN

Dean William Alderman of Beloit College claimed that "deaning in the '20s," whether at Purdue, Wisconsin, or on his small campus in Beloit, Wisconsin, was a difficult job for the heartiest of men. Public perceptions of college life in the 1920s, the era of flappers, bootleg liquor, and general licentiousness, Alderman claimed, was of a

> ...heteronomy of monstrosities—bobbed haired daughters of Satan in their early nicotine's; tardy sons of Hoyle who have an aversion for

> the hardy sons of toil; emulous coeds who indulge in such fatuous anachronisms as breaking the endurance record for the tango;...The daughters of culture have married the sons of prosperity and are sending their offspring to universities and colleges because it is fashionable, convenient and prudential.[16]

Alderman presided over a relatively small campus. At the University of Illinois, there were 7500 men compared to the 308 men at Beloit in 1933.[17] Even when the 244 women were added in, Beloit had but a fraction of the students at Illinois. But size was of little consequence. As a dean representing a small, private college campus, Alderman became quite active in the National Association of Deans of Men during the 1920s and 1930s. After serving for two years on the Executive Committee of the Association, Alderman became the NADM president-elect at the 1934 conference. Alderman presided over the meeting held on the campus of Louisiana State University in 1935.

William Alderman was typical of many of the deans who were members of the National Association of Deans of Men in the first half of the twentieth century. Born in Glouster, Ohio, in 1888, Alderman graduated from Ohio University in 1909. He received his masters degree from Hiram College in 1910, did graduate work at Harvard between 1912 and 1914, and earned his PhD in English Literature in 1920. Between 1914 and 1920, Alderman taught in the English Department while completing his degree.[18] He married his wife, Wilhelmina, also a graduate of Ohio University, in 1912. The Alderman's had four children, three girls, Barbara, 1914, Jane, 1918, Eleanor, 1920, and a son, William, Jr., in 1922.

Alderman noted that he had taught in the Student Army Training Corps (SATC) during World War I at Wisconsin. As Scott Goodnight (among others) noted, the SATC was a bad idea gone amok, so it can be assumed that Alderman knew or at least was acquainted with Goodnight, in his role as either a German instructor or Dean of Men at Wisconsin, possibly both. At the very least, they would have renewed their Wisconsin connections at the NADM meetings.

When Alderman became Dean of Men at Beloit in 1920, he also assumed the titles Dean of the College and Associate Professor of English. He stayed at Beloit until 1935 when, as he said in a letter to the President of Beloit, Irving Maurer, he had "an unsolicited offer from Miami University [OH] which from the financial point of view, is exceedingly attractive since it represents an increase in salary of from 40 to 45 percent over what I am getting here."[19] Alderman expressed a desire to remain in Beloit but Maurer informed him that he could not match the salary offered at Miami. Alderman became Dean of the College of Liberal Arts, Chair of the English Department and Professor of English at Miami in the Fall of 1935. He remained at Miami for the rest of his career. Because his new position did not include the title Dean of Men, he ceased his attendance at the NADM meetings at the conclusion of his NADM presidency. However, Alderman did seek out other professional associations. He had served a term as president of the Association of Presidents and Deans of Wisconsin Colleges, 1933–1934, and after his move to Ohio, a term as president of the Association of Ohio College Presidents and Deans, 1942–1943 and finally as president of the American Conference of Academic Deans from 1953 to 54.[20]

Alderman was very active among the deans of men and attempted to bring some of the newest developments among the deans to Beloit during his 15 years there. In 1928, he wrote to President Maurer of Beloit about the "personnel work" being done on other campuses. In particular, he noted, he had contacted "Dartmouth College, Oberlin College, Northwestern University, the Personnel Research Foundation, and the American Council on Education, et cetera (sic), have read Personnel Journals, and reports of Personnel Conventions, and have had conferences with those directing the work at Northwestern University."[21]

In his letter, Alderman explains that he had discussed the personnel needs at Beloit. After his consultation with "Doctor Howard at Northwestern," he felt reassured that little more was needed at Beloit beyond an effort to consolidate student records more carefully than had been done to date. An effort to inform students as to the value of more "vocational guidance" would be useful as well. Toward the close of his letter, Alderman expressed a concern echoed by many deans of men regarding the new "science" of the personnel movement. As he noted, "I am gratified that an increasingly large number of our graduates are going on to advanced work; but I am equally distressed to find that so many of them become bond-salesmen and bank clerks because they know of nothing better." Alderman can see the value of vocational guidance for these students as well as the value of consolidating information.

> when...an adviser or dean wants to get quickly a complete picture of all that the College knows about a certain student and of all that the student is doing and has done,—when any one of these emergencies arises, we realize how scattered and incomplete our data is. But at the same time," Vocational Guidance and Personnel Work as such can be

> too highly mechanized. I am skeptical of the system that would, upon the basis of a causal conference, classify a man as an "extrovert" or an "introvert" and tell him then and there that he is absolutely unfitted for certain work. We ought to do much, but it must be humanized rather than methodized.[22]

More pragmatic than many academics, Alderman, as a dean, could see the benefits of gathering information about students and consolidating it in a single location. The personnel movement in evidence at Northwestern and several other institutions was gaining some momentum. The logic of organizing information made sense in many respects. But the "mechanistic" nature of the personnel interview and the categorization of students on the basis of personality traits and characteristics troubled him as it did many deans.

After the disruptive storm of World War I created turmoil on campuses, the tempestuous student culture of the so-called Roaring '20s" descended on the same beleaguered faculty and administrators who were anxious for calm and order. When *Life* magazine sent photographers and reporters to cover stories of excess and salaciousness on college campuses, deans such as Alderman and others knew the hyperbole was overdone. Yet they were challenged by the changing nature of the students on their campuses. The increase in numbers in the post-war years was significant. In addition, women were now, on many campuses, near equal in quantity and in attitude. As women challenged the stereotype of the quiet, demure female by smoking, driving in cars, dancing, and the use of alcohol, the deans, both of men and women, were challenged anew.

But Dean Blayney of Carleton College in Minnesota noted what the public did not see and *Life* magazine did not report. From his perspective, "ideality and frugality still haunt college halls, and . . . students continue to bring their hopes and poems to their deans and professors." To respond appropriately "dean of men [must continue to] . . . be all things to all men."[23] In particular, Blayney argued, ". . . deans, especially deans of men, are quite universally considered as being primarily interested in business and administrative matter, rather than in scholastic matters." He urged that social and disciplinary issues were too great a focus of the dean's time and energy. "College deans . . . should form . . . a part of the shock-troops in the great struggle against inefficiency and superficiality in [American] higher education"[24]

At the same meeting of the NADM in 1928, Harold Speight, dean of men at Swarthmore suggested to his peers that, ". . . while there is no specific direction upon which we can expect general agreement in preparing men to serve as Deans and Advisers of Men, the greatest hope lies in the development of [the] profession. We should build up the profession through apprenticeships, and maintain...a list of young men...now in training."[25] Like their predecessors at Harvard and Brown, Blayney, Alderman, and Speight were faculty members who, unlike Dean Clark and Goodnight and others in larger institutions, kept their academic posts while assuming the administrative role of dean.

THOMAS BLAYNEY

Thomas Lindsey Blayney was a professor of German and Dean of the College at Carleton College in Minnesota. Blayney was well-travelled and had an intriguing history prior to his arrival at Carleton. Born in 1874 in Lebanon, Kentucky, Blayney was the son of a minister, Rev. J. McClusky Blayney, and his wife, Lucy Weisiger Lindsey Blayney. As a young man, Blayney earned his bachelors and masters degrees at Centre College in Danville, Kentucky, in 1894 and 1897, respectively. He completed his doctoral work in Germany at the University of Heidelberg in 1904. He returned to Centre College to teach for eight years and then became one of two full professors at Rice Institute (later University) in Houston, Texas, in 1912.[26]

During World War I, Blayney served as a staff intelligence officer with British, French, and American armies, and held the ranks of major and lieutenant colonel. After the Armistice, he was sent by the American Peace Commission into Germany to hold hearings and report on economic conditions. He returned to Rice, where he was prominent in speaking out against the Ku Klux Klan. He later served for two years as president of Texas State College for Women before moving to Carleton in 1926 as Professor of German (until 1946) and Dean of the College (until 1945).[27] He died March 13, 1971, at his home at Marine on St. Croix.

HAROLD SPEIGHT

Harold Speight served as Dean of Men in 1933 and then became Dean of the College at Swarthmore. A Unitarian minister, Speight was born in Bradford, England, in 1887. He earned his M.A. from the University of Aberdeen in 1908 and began his academic career at Aberdeen as an Assistant in Philosophy. He then spent two years as a fellow in Exeter College at Oxford before entering the ministry.

As a minister, he served congregations in London, Victoria, British Columbia, Berkeley, California, and finally at King's Chapel in Boston. In 1927, he returned to academia and took a position at Dartmouth, where he was Professor of Philosophy. He later became Professor of Biography and chair of the department. In 1933, he took the position as Dean of Men at Swarthmore. He then moved to deanships at Cornell and Elmira College and served as acting President at Sarah Lawrence during World War II. He returned to Dartmouth after his retirement and lived there for many years before moving back to Victoria, B.C., where he died in 1975.[28]

Speight was not only a Unitarian minister and professor of philosophy and biography, he authored *The Life and Writing of John Bunyan* in 1928 as well as a set of biographies titled *Creative Lives.* He published some of his sermons from Kings Chapel. Speight was an active member of the Peace Conference at Swarthmore, a group that campaigned actively against U.S. entry into World War II.

Harold Speight attended the 1935 meeting of the National Association of Deans of Men held at Louisiana State University and gave a paper titled "Stimulating Intellectual Activity." In an interesting combination of his several roles, he was still Dean of Men at Swarthmore when he gave a speech at the University of California, Berkeley, for the 1396 Foerester Lecture on the Immortality of the Soul. His speech was titled, appropriately enough, "The Immortality of the Soul."[29] He was a rare dean of men, indeed.

Herbert Hawkes

"We believe that the college must be as concerned as is the parent with the question of what can be done to justify a student's spending four of the most determining years of his life within its walls. If it is not the right college for this student or if he is frustrated in unforeseen relationships, and the college does not help him to find himself, he cannot return his degree to the shop and get his four years back."[30] So said Dean Herbert Hawkes and his wife, Anna Hawkes, in *Through A Dean's Open Door*, published posthumously after Hawkes sudden death in May, 1943.[31]

Herbert Hawkes was a mathematician by training. He entered Yale at 20 and completed his undergraduate work in 1896 at the age of 25. He completed his doctoral work at Yale in 1900, married, and briefly studied in Germany, all in the same span of time. Returning to Yale as an assistant professor, he stayed until 1910, when he moved to Columbia.[32] Hawkes authored seven textbooks in mathematics as well as four books on higher education and the use of achievement tests.[33] One of his biographers, Jo Ann Fley, contends that Hawkes felt his administrative duties at Yale caused him to be passed over for promotion. Frustrated, Hawkes moved to Columbia, where he accepted a full professorship and made a personal commitment to only pursue research and teaching.[34] Despite his best efforts to avoid administration, Hawkes was unable to repress his administrative talents, which seemed to bubble to the surface, and he was soon taking on the administrative tasks at Columbia, such as oversight of all undergraduate education in mathematics. His easy and likeable manner with students and faculty alike made him a natural choice to be Dean of the College, a role he assumed at Columbia in 1917.

Hawkes took up the responsibilities of his new position with a commitment to do what was best for the students. As a mathematician, Hawkes was comfortable with the enumeration of student characteristics and traits and the quantification of various measurements and tests in assessing college students. He quickly adopted the basic precepts of the new "personnel movement" that emerged in the early decades of the twentieth century. The personnel movement, which became known as the "student personnel movement" when it was applied to college students, encouraged extensive record-keeping, the use of tests and measurements of various aspects of personality, and the accumulation of records as a means of providing advice and direction to students and to their families.

Not only did Hawkes begin to use the measurement techniques of the "student personnel movement" on his own campus at Columbia, he also became the chair of the Committee on Personnel Methods in 1925 at the instigation of George F. Zook, president of the American Council on Education (ACE). Hawkes became an intrepid advocate of the personnel movement and eventually chaired the ACE committee on Student Welfare which, in 1937, published the now iconic brochure, *The Student Personnel Point of View*, which set the stage for a new era in student services.[35] Hawkes and Zook maintained a 20-year relationship professionally and personally. According to Zook, Hawkes chairmanship of the Measurement and Guidance Committee for ACE guided much of the progression of the student personnel movement.[36]

WALTER DILL SCOTT

Although he was never a Dean of Men, Walter Dill Scott had tremendous impact on the deans of men and on higher education in the twentieth century in general. One of the earliest adoptions of the new social science of psychology to a college campus was at Northwestern University where Walter Dill Scott, an industrial psychologist by training, applied the relatively new field of personnel psychology to the college campus. Born shortly after the Civil War ended in 1869, Scott grew up in rural Illinois in a farm family of five children. He attended Illinois State University for two years and then matriculated at Northwestern University. He completed his degree in 1895. Scott's childhood dream was to be a teacher. However, his goal in college matured into assuming the presidency of a university in China. Because most universities there were sponsored by religious groups, Scott enrolled at McCormick Theological Seminary and earned a second degree in 1898.[37] When his path to a presidency was not forthcoming, Scott followed his growing interest in psychology to the University of Leipzig.[38] Scott's PhD was earned in psychology and educational administration in 1900. His wife of several years, Anna, earned her PhD in philology and art from the University of Halle at the same time.

At Leipzig, Scott studied under Wilhelm Wundt, a pioneer in psychology who "separated" the new field from its origins in philosophy. Scott returned to Northwestern as an instructor in psychology and pedagogy in 1900, became a professor in psychology in 1907, and department chair in 1909. He wrote two books, *The Theory of Advertising* (1903) and *The Psychology of Advertising* (1908), in which he explored the applications of psychology to business and industry. Scott was instrumental in developing tests and other measurements for use in business and industry, and called his work "personnel psychology," as it was used to assess the personnel or people working in a business setting.[39] Scott was given leave from Northwestern to head the Bureau of Salesmanship at Carnegie Institute of Technology. When World War I broke out, Scott and a team of psychologists working with him volunteered their services to the military, in particular to help with officer selection.

The Army agreed and Scott, along with his colleagues, helped use their batteries of tests and measurements to develop procedures for officer selection. Scott's techniques proved very successful. He was promoted to colonel and awarded a Distinguished Service Medal before he left the Army. After the war ended, Scott was elected President of the American Psychological Association in 1919. In the same year, he started a very successful company, Scott Company Engineers and Consultants in Industrial Personnel. With offices in Chicago, Philadelphia, and Dayton, OH, the new company provided services to over 40 clients in the first year and was on track to be very successful.[40] However, when his alma mater, Northwestern University, invited him to return to be the new president of the university, he agreed and assumed the presidency in 1920.[41]

Scott's rapid rise through the ranks of the military and the business community encouraged him to apply his new techniques in his newest setting, the college campus. Soon after his appointment, Scott began to instruct his staff in the application of the "personnel movement" in their work with college students. Re-naming the technique, the "student personnel movement," Scott recruited two earnest protégés, L. B. Hopkins and Esther Lloyd-Jones.[42]

L. B. Hopkins, one of Walter Dill Scott's strongest colleagues in the Bureau of Salesmanship and a member of the newly formed Scott Company, had developed a battery of tests for the Wilson Sporting Goods Company. The program was very successful in identifying personnel characteristics and attributes. Scott lobbied Hopkins and Wilson for permission to adapt the program for the college campus. The Wilson Company not only granted their permission, they agreed to help fund the effort. In turn, Scott made Hopkins the new Director of Personnel Services at Northwestern.

The connections between Scott, Hopkins, a young graduate student, Esther Lloyd-Jones at Northwestern, and Herbert Hawkes at Columbia are numerous. One of the first links came when Hawkes, in his role as chair of the American Council on Education's Committee on Personnel Methods, asked L. B. Hopkins at Northwestern to conduct a study to determine what institutions across the United States were doing to assist in the development and advisement of individual students. In short, he wanted Hopkins to determine the number of institutions that were using the new "personnel methods" endorsed by Scott, Lloyd-Jones, and Hopkins as well as the methods they were using. Hopkins published the results of his study in the May 1926 issue of *The Educational Record,* the journal of the American Council of Education. Hopkins also presented his findings to various groups, including the National Association of Deans of Men at their national conference in 1926.[43]

In the midst of this activity, Hopkins assumed the presidency of Wabash College, a small, single-sex liberal arts college for men in Crawfordsville, IN. One of Hawkes' books, *Five College Plans:Columbia, Harvard, Swarthmore, Wabash, and Chicago,* published in 1931, examined five colleges, including Wabash. The comparison of the small, rural college in Indiana (Wabash) to three heavyweights of higher education (Harvard, Chicago,

Columbia) is interesting, to say the least. Wabash was a somewhat unique institution but it is too coincidental to ignore Hopkins move to Wabash and the inclusion of Wabash College in Hawkes' book.

Esther Lloyd-Jones, another Scott protégé like Hopkins, completed her undergraduate degree and then began her career as a graduate student in psychology at Northwestern. She took courses with Scott and became acquainted with Hopkins as well. Her interest in personnel work was piqued and her doctoral dissertation detailed the development of the personnel methods at her alma mater. She later published it as a book, *Student Personnel Work at Northwestern*, in 1929. Scott purchased copies of her book when it was published and distributed them across campus, sending them to the Board of Trustees, faculty, and even students.[44] Lloyd-Jones eventually left Northwestern in 19 to become a faculty member at Teachers College, Columbia University. Lloyd-Jones joined a strong group of women already at Teachers College in the Counseling and Guidance Department, a group that included Sarah Sturtevant and Ruth Strang, early and vibrant research-focused pioneers in the personnel and guidance field. In 1937, Lloyd-Jones, and Hawkes would work with each other on a report from Hawkes titled, "The Student Personnel Point of View."

The relationships within the American Council of Education were even more extensive than just Zook, Hopkins, Lloyd-Jones, and Hawkes. Zook also included W.H. Cowley, of the Bureau of Educational Research at The Ohio State University. Cowley earned his undergraduate degree at Dartmouth and completed his doctoral work at the University of Chicago after a brief research stint at Bell Laboratories.[45] Cowley's work at Ohio State included "the improvement of student personnel services there and nationally"[46] Cowley was an Assistant Editor on the *Journal of Higher Education* when it first appeared in 1930, working with the Director of the Bureau and Editor of the Journal, W.W. Charters. Cowley also became a close colleague of George Zook, and the work of the ACE. (Zook, a historian, had been president of the University of Akron before he took on the role of President of the ACE and, briefly, U. S. Commissioner of Education in the 1930's—making the Ohio connections intriguing.)

Over time, as the men listed above embraced many, if not most of the concepts and ideas of the student personnel movement, the practical application of Walter Dill Scott's ideas began to appear on more and more campuses. In 1936, the advance of the "personnel movement" had been sufficiently engaged at Northwestern that Scott eliminated the positions of Dean of Men and Dean of Women and replaced them with a Board of Personnel Administration that combined counseling services (for both undergraduates and graduates) with housing, financial aid, admissions, student records, and placement services.[47] Soon after this development, the American Council on Education's Committee on Student Welfare issued their brochure, "The Student Personnel Point of View," which became the rallying cry for change like that at Northwestern to occur across the United States. The members of the committee, under the purview of the ACE and George Zook, were becoming very familiar as they included Herbert Hawkes who chaired the committee, Esther Lloyd-Jones, L.B. Hopkins, and W. H. Cowley.

Within a year of the ACE committee report, W. H. Cowley addressed the National Association of Deans of Men annual meeting in 1937. The title of his speech was "The Disappearing Deans of Men."[48] Although the deans of men would not disappear for some time to come, the warning shot had been fired. Those deans who attended the NADAM (by that time, the name of the association had been changed to include "advisers of men") were certainly challenged by the directness of Cowley's assertions and the tides of change that he forecast were coming.

The deans of men in private colleges and universities were welcomed into the NADM (and NADAM) as openly as those from public institutions, but their roles and responsibilities were different. In many cases, the deans in private institutions had to, of necessity, wear more hats. They were often not just the dean of men but also deans of their colleges and more often than not, held faculty rank and taught on a regular basis. Some of the earliest men to accept the role of dean of men, such as Briggs and Meiklejohn, were never members of the national association. Briggs preceded the association by many years and Meiklejohn was destined to positions far beyond the deanship.

Those men who did serve as deans in private colleges or universities were often placed in a far more intimate roles with other faculty and with students than were deans such as Clark or Goodnight or other deans in the larger, public institutions. Clark thrived on the large campus setting of Illinois and delighted in designing mechanisms for the control and maintenance of complex issues and problems such as new student orientation, an infirmary to address student health concerns, and the registration of student groups and organizations through his office. As Dean, he had a staff larger than the president of the university and the constant flow of students through his office was daunting.

On the private college campus, the relationships were more intimate and personal. The need for mechanisms such as Clark put into place were non-existent. Yet at the same time, it was often the deans in private schools—Hawkes, Alderman and others—who were eager to consider or even, in Hawkes' case, embrace the new technology of the personnel movement. Though Clark and others were aware of the personnel psychology approach to work with students, much of the movement occurred after their terms had ended. Certainly that was the case with Thomas Arkle Clark, whose retirement and death occurred early in the 1930s. Others, like Scott Goodnight, adapted as much as they could, but even Goodnight retired before the personnel movement changed everything, including the NADM.

5

Francis F. Bradshaw: A Southern Student Personnel Pioneer

Image 3 Francis F. Bradshaw, Dean at the University of North Carolina-Chapel Hill from the early 1920s through the 1940s. Courtesy of the American College Personnel Association Archives (MS-319); Center for Archival Collections, Bowling Green State University, Bowling Green, Ohio.

At the 1931 meeting of the National Association of Deans of Men in Gatlinburg, Tennessee, a spirited discussion about the preparation needed to be a dean of men took place. Most of the deans of men expressed the belief that the best preparation to be a dean of men was to be "born" to the position. In short, the assembled deans believed that specific training or even graduate education would do little to prepare a man to be a dean if he didn't have the right temperament for the job.[1]

Speaking at the meeting, Joseph Bursley, dean of men at the University of Michigan, summarized the NADM theme with his remark, "I am afraid that I am not in sympathy with the idea of any course of training for the position of Dean of Men. The best and most successful Deans of Men are born and not made."[2] However, Bursley continued,

> There is one place where I believe preparedness is absolutely essential to the success of a dean of men—that is in the selection of a wife. The very best preparation he can have for his work is to marry the right woman. If she is the right kind, a dean's wife does just as much to earn his salary as he does, and if she is not, he might as well quit before he starts.[3]

The lone dissenting voice challenging this idea belonged to Francis F. Bradshaw, dean of students at the University of North Carolina. An active participant in the new "personnel movement," Bradshaw made his feelings known.

> In our deanly world we are not saved by any training processes whatsoever. This makes it a little difficult to talk about the dean of men's preparation for his work. The last speaker, however, gave me some hope through his statement that while "deans were born and not made," they might be made better by preparation. The deanship stands to some extent at a fork in the road.... whether we are to be solely campus disciplinarians or whether we are to be administrative coordinators of the whole individual student and...of group life of students. The discovery of the genuineness and permanence of individual differences by modern psychological science, the rapid expansion of our campus communities,...set up a demand for a[n] administrative...office resting on these fundamental points of view.[4]

Bradshaw's comments countered the dominant point of view at the NADM meetings. But then, Bradshaw was not like many of the other deans present. He had been Dean of Students at North

Carolina, his alma mater, since 1920. Born in Columbia, South Carolina, Bradshaw had earned his A.B. at North Carolina in philosophy in 1916. A veteran of World War I military service, Bradshaw completed his PhD in psychology in 1930 at Columbia by taking one year of leave from the University of North Carolina in 1925–1926 and by working at Columbia in the summers of 1925–1929 teaching psychology while completing his course work and dissertation. He was elected to serve as the president of the National Association of Appointment Secretaries (NAAS) in 1928 and re-elected to a second term in 1929.[5] The NAAS would eventually, after several more name changes, become the American College Personnel Association.

Most deans in attendance at the NADM in 1931 followed Bursley's notion that good deans were born, not made. They resented the idea that the personnel movement, which Bradshaw strongly supported, could be superior to the inherent skills of a good dean. As Stanley Coulter of Purdue described it, "the first time I met with the Deans of Men was at Illinois [1920] we discussed the same problems as now. We had little of mechanical devices for solving these problems. Today [1928] we have so surrounded ourselves with mechanical records that we may have ceased being personalities and have become machines."[6]

"Mechanical records" were a part of the public trail left by the personnel movement. To those who embraced the student personnel movement, "mechanical records" were snippets of an individual student's biography, bits of data collected over time by administrators in the college or university. Appointment cards or personnel records were used to collect information on each student. Information kept included family size, birth order of siblings, occupations of parents, high school attended, and the like. As the student progressed, college records such as courses taken, grades, social activities, and more would be added.

Eventually, a student record was to be accumulated from the beginning of the freshman year and throughout the student's entire academic career, advising notes, comments from faculty, and others would be added. By keeping these cards on each student in a central office, the collected information could then be used to offer advice and guidance to students as they progressed through their program of study and prior to graduation. By consulting the student's personnel record, advisors and faculty alike could direct the student toward the appropriate major or field of study and ultimately, towards a career most suitable for them and their unique characteristics and predilections.[7]

As delineated by Walter Dill Scott, L. B. Hopkins, and Esther Lloyd-Jones at Northwestern, the personnel movement when applied to college campuses became the "student personnel movement." It was this application of record keeping and the measurement of personality traits and abilities that so intrigued F.F. Bradshaw when he became the Dean of Students at the University of North Carolina. Perhaps he himself had been exposed to the "personnel movement during his brief stint in the military during World War I. Walter Dill Scott's first real application of the personnel psychology techniques were used to determine officer candidates in Ft. Myers, NJ and other locations."[8] At the very least, Bradshaw might easily have learned about the application of personnel psychology from other officers.

The "personnel movement" naturally grew out of the social efficiency trend of the late nineteenth and early twentieth centuries. As a branch of social efficiency, the personnel movement and its step-child, the student personnel movement, took many different strategies, each intended to increase the efficiency of human skills and talents by mapping individuals according to their abilities. Frank Parsons, often referred to as the "father of vocational guidance," was one of the key players in the movement.[9] Frank Taylor, a mechnical engineer and the father of the efficiency movement in industry, and Frank Gilbrith, a proponent of time and motion studies, were other significant figures in social efficiency who are still remembered for their work studies.

Bradshaw's encouragement of vocational guidance and psychological testing, interviews, record keeping, and "objective" measures of personality put him at odds with the other deans in the NADM. As noted, many of Bradshaw's peers in the NADM argued vehemently against the use of guidance and counseling techniques but also against graduate training to be a dean. F. F. Bradshaw was one of the few deans of men in office from 1920 to 1930 to follow a more contemporary track in his work with students. Bradshaw can and should be recognized as one of the first deans of men to understand and embrace the professional "sea change" that eventually led to the establishment of modern student affairs work.

When he was appointed to be Dean of Students at the University of North Carolina in 1920, Bradshaw was the second person to hold the position. Frank P. Graham had been the first dean but only held the office from its creation after World War I in 1919 to 1921, when he resigned to return to teaching in the history department.[10] Women were admitted to the University of North Carolina in 1898 (only to the upper classes) but even by 1919, only 47 women were enrolled. Nonetheless, these few women were sufficient to prompt the administration to appoint an advisor to women. In fact, Mrs. T.W. Lingle was named Adviser to Women in 1917, two years before Graham was named the first dean of students.[11] So though Graham and then Bradshaw both held the title Dean of Students, it is clear that most of their work was focused on male students.

Although Frank Graham left the position of Dean of Students to return to the History department, he was back in the university administration within several years. Frank Porter Graham became the president of the University of North Carolina in 1930. Graham has often been given credit for the far-sighted and energetic expansion of the university from a small, Southern school with 2500 students and 250 professors to an exemplar of intellectual and scholarly activity and a member of the prestigious Association of American Universities (AAU).[12] Graham was a gifted administrator and a staunch defender of his faculty and students. Although he stopped at the actual admission of African Americans to the university, he helped his faculty, especially those in sociology, history, and theatre, host African American artists such as Langston Hughes, in Chapel Hill in the 1930s. Graham was an "exceedingly popular Mr. Chips" on campus, according to John Edgerton, and the university "blossomed like a desert flower" under his leadership.[13]

Frances F. Bradshaw's career initially followed the same path as his predecessor, Frank Graham. Like Graham, Bradshaw was an ambitious young man and a graduate of the university. He became the YMCA secretary, just as Graham had, and pursued a graduate degree in philosophy. When he replaced Graham as Dean of Students in 1920–21, Bradshaw was a young man of 27 and shared the same idealistic traits as Graham. But unlike Graham, Bradshaw did not ascend to a college presidency. Instead, he became very involved in the student personnel movement and a leader on the national scene through his involvement in the American Council on Education, the National Association of Appointment Secretaries, and the National Association of Deans of Men.

Bradshaw's involvement in the university following his undergraduate career and military service started innocently enough. Many campuses had been converted into quasi-military encampments by the Student Army Training Corps (SATC) as a part of the national mobilization for World War I. The war drew many young men from colleges and universities into the officer corps by creating a shortened active duty program on campuses. Land grant colleges and universities, especially, were required to provide compulsory military training, so a rapid turnaround of students into officers was not difficult.

F. F. Bradshaw himself had been commissioned as a second lieutenant and served in the Field Artillery from 1918 to 1920. He returned to the North Carolina campus as General Secretary of the campus-based Young Men's Christian Association (YMCA), from 1916 to 1918.[14]

After the war ended, however, the dismantling of the military presence became a primary concern on many campuses. The University of North Carolina was no exception. Students, faculty, and administrators were eager to restore an academic climate on campus and to remove the military theme created by the SATC. Scott Goodnight of Wisconsin, as did other deans, bemoaned the impact the SATC had on the college campus. It was a "nightmare," Goodnight lamented, as the military presence on campus altered the campus culture in significant ways.[15] Deans of men on other campuses found the anti-collegiate impact of the military rules and regulations very disruptive and an anathema that worked against their own administrative work.

Within months of assuming the office of Dean of Students, Bradshaw wrote to other universities enrolling more than 1,500 students to ask about their dean of students office and the services they provided. He personally visited Harvard, Yale, and Columbia in his first year as dean (1921–22) to observe and visit with other deans.[16] Bradshaw also attended the second meeting of the newly formed National Association of Deans of Men held in Champaign, Illinois. A few years later, Bradshaw hosted the seventh meeting of the NADM in Chapel Hill in April 1925.[17] He also served as secretary for the NADM for a three-year period beginning in that same year, 1925, and continued through 1927.[18]

Academically, Bradshaw added to his duties as Dean by teaching a course on ethics in the Philosophy department at North Carolina. Then, in an unusual step for a dean of men in the 1920s, Bradshaw enrolled as a graduate student in psychology at Columbia University in the summer of 1925 and stayed through the academic year as well as the summer of 1926. He earned his A.M. degree in the Fall of 1926. He returned to Columbia in the summer of 1927 for more course work, and taught summer school while taking classes and working on his dissertation in the summers of 1928 and 1929. He earned his PhD in psychology in December 1930. His dissertation was titled, "The American Council on Education Rating Scale: Its Reliability, Valdity, and Use."[19]

In addition to his graduate courses at Columbia, and his role as Dean of Students at North Carolina, and his involvement with the NADM, Bradshaw assumed the presidency of the National Association of Appointment Secretaries from 1928 to 1930. Appointment secretaries were often in dual roles of offering advice and direction to students regarding careers and occupations upon graduation as well as helping students find employment while in school. Beginning in the 1920s they were also tasked with the compilation of personnel records for students beginning in the 1920's. As the association grew and became more diverse in both application and direction, the NAAS became better known as the American College Personnel Association.[20]

Even before completing his Columbia graduate courses, Bradshaw vigorously challenged the status quo of student life at North Carolina. Soon after his appointment as Dean in 1921, Bradshaw separated himself from direct responsibility for the administration of discipline on campus. Although many deans of men often found discipline to be a handicap in their work with students, they did it anyway. In some respects, these were the onerous duties that had framed the need for deans of men and to a lesser degree, women, from the beginning.

But discipline duties saddled deans of men with a negative image they disliked and sought to minimize. Discipline for many of the deans was a small part of their job. Some deans, such as Thomas Arkle Clark, seemed to embrace the role, he even wrote a book on the topic, *Discipline and the Derelict* (1921). Howard Hawkes also saw discipline as an opportunity to advise students, albeit from a different angle. But many deans agreed with Robert Rienow, at the University of Iowa, who declared, "For thirteen years, I kidded myself that I was not a disciplinarian. But in the last few years, I have learned that I was [always] pulling the chestnuts out of the fire for someone else"[21]

Instead of assuming the traditional role of chief campus disciplinarian as a part of his duties as dean of students, Bradshaw delegated discipline to others. He developed a new Committee on Student Life which he chaired. He also organized four sub-committees under Student Life including the areas of: supervision, coordination, personnel work, and investigation. Student Life became the key focus of Bradshaw's move to a personnel movement on campus at Chapel Hill, comparable to other innovative programs such as Northwestern University in Illinois.[22]

In the same year, Bradshaw established a Freshmen Counseling system with 50 members of the faculty assigned to work specifically with new freshmen. He organized the Publications Union, which was to promote student activities in various areas, and re-organized the chapel system, which was required for all students. Bradshaw also sought approval to initiate a new vocational guidance program for all students.

By 1922–1923, his second year in office, Bradshaw was still making changes and creating new committees. He established a new Bureau of Vocational Guidance on campus. He also streamlined the membership process for fraternities, instituted intramural athletics for the residence halls in the form of tag football, and further expanded the student activities and organizations on campus. He added a $5.50 student fee to support the work of the Publications Committee and even oversaw the installation of phones in campus residence halls.[23]

In his fourth year as dean, Bradshaw had come close to his goal of completely restructuring the culture of campus life for students. He had established new systems of advising, student activities, and instituted student fees. He also chaired various committees responsible for student life and established the university *Bulletin*.[24]

Acting more as a personnel worker than a traditional dean of men, Bradshaw directed freshmen counseling and vocational guidance himself. He chaired the Faculty Committee on Vocational Guidance and Student Employment, administered student loans, and made referrals to the Personnel Research Bureau for Mental Hygiene. Over the next several years, Bradshaw worked extensively to expand the role of vocational guidance and student personnel in directing the careers and academic work of students at North Carolina.[25]

F. F. Bradshaw had clearly embraced the new personnel movement that was slowly but surely making its way onto the college campus. The Report of the U.S. Commissioner of Education for 1916–1918 (1919) noted in particular the work of the Intercollegiate Intelligence Bureau which had, under the direction of Dean William McClellan of the University of Pennsylvania, "prepared personnel records of those members of…student and alumni bodies and faculties"[26] as a part of the war effort. Although it was not embraced on many campuses, the advent of the new student personnel movement, as it evolved on college campuses, began to change the administration of student life.

Bradshaw's early work with the local Young Men's Christian Association (YMCA) might have been a source of his interest in vocational guidance and, by affiliation, to the personnel movement. According to Savickas and Baker, the YMCA movement often had chapters in large cities but also on or near many college campuses. In the early twentieth century, the YMCAs had changed their focus from working with the destitute (alcoholics and discharged convicts) to prevention. A large part of the effort was to help young men find employment.[27]

Some of this early effort at education and placement was fueled by C.C. Robinson's program, *Find Yourself*, published in 1912 and again in 1922. The program was offered "to every young boy who entered a YMCA program." Further, as noted by Savickas and Baker, the focus on vocational guidance services in the YMCAs "reached their zenith in the 1920's and 1930's"[28] The first course in vocational guidance was offered at the Boston YMCA in 1908 and was to have been taught by Frank Parsons himself. Unfortunately, Parsons died before the class began and a friend taught it instead. Later, the course and concept moved to Harvard as the Bureau of Vocational Guidance. Harvard first offered a course in the field in 1911. Similar courses were developed at the University of Chicago (1912) and Columbia (1913).[29]

When World War I began, the vocational guidance movement and the psychologists who were active in the field promoted the benefits of personality tests and the like to help the war effort. As has been noted elsewhere, most campus-based applications can be traced back to the influence of Walter Dill Scott and two of his protégés, L.B. Hopkins and Esther Lloyd-Jones.[30] The student-focused personnel movement was in many ways related to the new vocational guidance movement. The two schools were both keenly focused on the early investigation of an individual's goals and aspirations. There was also great interest in cataloguing skills, traits, abilities, and even movement and physical skills. The "father of vocational guidance," Frank Parsons, and others, directed much of their research and writing towards the general labor force. But Walter Dill Scott used his psychology training to focus on first the military, then industry, and finally, college students.

Walter Dill Scott was an early advocate of personnel sciences that advocated the measurement of abilities, aptitudes, and talents of individuals to determine their vocational path. By instituting a battery of tests and individual interviews to match individuals with appropriate jobs, Scott believed that both the individual and society would be made more efficient and more satisfied. Scott found ample opportunity to test out his ideas. He first applied his theories to industrial settings in the early 1900's; the assignment of factory employees to jobs was determined by aptitude testing and individual interviews. Scott, along with others, then applied the same principles to draftees during World War I.[31]

When World War I ended, Scott assumed the presidency of Northwestern University, where he used the "scientific principles" of the personnel movement with students at Northwestern and the "student personnel" movement began. L.B. Hopkins and Esther Lloyd-Jones were staff members of Scott's at Northwestern. Hopkins served as Director of Personnel at Northwestern and authored a monograph,

Personnel Procedure in Education (1926), published by the American Council on Education (ACE). Hopkins' monograph was one of the earliest accounts of the personnel approach used with college students. Esther Lloyd-Jones was a doctoral student and administrator for women at Northwestern. She wrote *Student Personnel Work at Northwestern University* (1929). Lloyd-Jones later joined the faculty at Teachers College, Columbia.[32]

Walter Dill Scott was an early proponent and author of the personnel movement. Scott advocated the measurement of abilities, aptitudes, and talents of individuals to determine a vocational path for the future. Believing the basic tenet of vocational guidance theory to be true, Scott sought the means to capture the necessary information to advise, some would argue categorize, human beings into appropriate work based on their intelligence, natural traits and skills, aptitudes, and stated preferences. From his own experience in developing a classification for the Army in World War I as well as the increasing emphasis on the use of psychology, a social science that could "map" the characteristics of the individual person, Scott was able to attach himself and his ideas to the growing interest in "appointment bureaus" and vocational training for the general population.

A conference hosted by the National Research Council in 1925, titled appropriately the Conference on Vocational Guidance, included representatives from 14 colleges and universities, as well as others from within the vocational guidance movement. Five members from the assembled group were selected to conduct a survey of personnel methods and to find a source to fund the survey.[33] This group and their work were subsequently adopted by the American Council on Education (ACE), where the committee was re-named the Central Committee on Personnel Procedures and chaired by Herbert E. Hawkes from Columbia. Other members of the committee included W.W. Charters from Ohio State, and L.B. Hopkins, who was still at Northwestern, before he became president at Wabash College, among others.

The new ACE committee secured $7500 from the Rockefeller Foundation to conduct their survey of personnel methods around the country at colleges and universities. The survey was conducted by Hopkins and issued as an ACE report in the October 1926 issue of the *Educational Record* and published separately as well.[34] The conclusions were that though personnel methods were in wide use (or at least reportedly so), the implementation of the methods showed "great variation, hesitation, and fumbling in actual procedures."[35]

To remedy these concerns, another meeting of the Central Committee was held on July 1 and 2, 1927. Five subcommittees were organized, each with a specific task related to the personnel movement and its implementation. There was a committee for personal record cards, one for achievement tests, one assigned to personality measurement, and one for occupational information. A fifth committee to address personal development was added a year later.[36] F. F. Bradshaw was appointed to serve on the Personality Measurement sub-committee chaired by D. A. Robinson of the ACE. It should be noted that the work of the committees was directed at secondary (high) schools as well as colleges and universities. The idea inherent in the inclusion of secondary schools was that the "personnel" effort should begin as early as possible. Students who were given guidance and direction in high school would benefit even more from such direction in college. In addition, although the committees were clearly dominated by men, four women were included in the group.[37] The five sub-committees were funded by an additional $20,000 over three years from the Rockefeller Foundation.

In 1931, a national survey of land-grant colleges and universities published by the Office of the U.S. Commissioner of Education included a chapter on "Personnel Service" in higher education, which explained the role of personnel services on the college campus. In some respects, this section of the report reflected the work of the ACE committees through 1930.

> The term "personnel service" has been carried over into the colleges from industry, where it came into prominence just after the World War [I]. Industry uses personnel work in selecting, teaching, lessening turnover, conserving health, and providing recreations for its workers.
>
> When colleges took over this type of work, however, the emphasis was totally changed. The fundamental aim in personnel work in the college is that of service to the student as an individual, and its entire organization centers around this aim.
>
> Those apparently who use the material gathered in psychological tests most frequently are the deans of the colleges, the deans of men and the deans of women, the faculty and administrative officers actually concerned with the individual student, prospective employers and industrial representatives. The colleges have for years been obtaining and filing away vast numbers of records, a source for limitless research, that, rightly used, might throw much light on many of their unsolved problems.[38]

By these standards, Francis F. Bradshaw was well ahead of his time at North Carolina. He had instituted a clear plan of personnel procedures by the mid-1920s, highlighted by his development of a vocational guidance board, the so-called investigation process, and the involvement of a team of 50 faculty to advise and work with the new freshmen class each year.[39]

In this light, however, it is not surprising that Bradshaw did not find a welcome reception when he reported on his innovations at the national meetings of the deans of men. At the 1931 NADM session, Bradshaw found himself on the periphery of the deans' sentiments. Many of the pioneering deans of men, most notably Thomas Arkle Clark of Illinois and Stanley Coulter of Purdue, had argued against the formality and artificiality of the scientific method of counseling students. These derogatory terms, "formality and artificiality" referred specifically to the personnel methods Bradshaw had employed at Chapel Hill. They were also opposed to the notion of graduate study, an attitude and bias that became a hallmark of the dean of men's meetings from the 1920s through the 1940s.

The deans of men were honestly convinced that the formality and abstraction of applying scientific processes to young students would exclude the human face of the administration and especially the dean of men. Many of the early deans of men, including L.B. Briggs and others, came to their jobs by way of personality and a natural proclivity for working with young people. They relied more on intuition and perception than scientific principles to achieve their successes.

Many of the early deans believed that they could train others in the same manner. Through self-selecting the right sort of young men, the deans of the NADM believed they could continue to promote the position of dean of men by passing on the trade through apprenticeships to carefully selected heirs. There was also some resentment toward the deans of women inherent in the deans' stance. The rapid growth of interest in vocational guidance just prior to and especially after World War I was a boon to the work of the deans of women. Long an ongoing concern of both the deans of women and the Association of Collegiate Alumnae, vocational training and bureaus of occupation came into greater prominence in major cities and on the college campus in the 1920s. Deans of women had also embraced the notion of graduate study in higher education early on.

Not surprisingly, Bradshaw took his energies elsewhere. He was absent for several years at the NADM meetings after 1931 (*Secretarial Notes of the National Association for Deans of Men, 1931–1940*). He became very active in the development of the new American College Personnel Association, the National Vocational Guidance Association, and later, an umbrella organization that attempted to bring the various counseling and guidance groups together under one name, the American Council of Personnel and Guidance Associations.[40]

Eventually, Bradshaw returned to the NADM. In 1939, he attended the NADM meeting held in Roanoke, Virginia. Once again, he addressed the assembled deans on the matter of the future of the profession of dean of men. In particular, Bradshaw stated

> I cannot help but feel the problem we are facing in deciding whether or not we are willing to be a staff office or a general executive office. I am convinced that the personnel point of view in the general personnel program is here to stay and expand for a while, and it seems to me that in that situation, the deans of men may either be specialists in discipline, social education or whatever you may choose to call it, or the causes of a more general type of training and personality may develop him into a general executive in which case, in the coordinated set-up, it seems to me that he might easily move in the direction of the central personnel coordination. But, he would have to make his viewpoint and training as broad as theirs is, in expecting to coordinate.
>
> If we go in that general direction, it is essential to associate more with other people in personnel guidance fields.... If we are to coordinate a general feeling, we must have more or less continued association with members of that same feeling.... I think we should consider a meeting at least every other year with the personnel and guidance associations.[41]

Grudgingly, the deans of men agreed to send representatives to meet with the Council of Personnel and Guidance Associations (CPGA) and address those issues of common concern. No doubt the deans were persuaded in part by the speech given only two years earlier in 1937 by W.H. Cowley of the Bureau of Educational Research at Ohio State. Cowley titled his presentation "The Disappearing Dean of Men."[42] Cowley had chaired the American Council on Education subcommittee on student welfare. In 1937, the same committee had authored a monograph, *The Student Personnel Point of View* (ACE, 1937), a small document that was delivered nationally to presidents and faculty at most colleges and universities.

Clearly the monograph delineated the direction of student services for the coming decades and followed the student personnel course Bradshaw had himself implemented at the University of North Carolina in the early 1920s. When Cowley discussed the monograph at the NADM meeting, he could not help but chide the deans to look closely at the future of higher education and student services in particular with the title of his speech. As noted in many arenas, the personnel movement meant a focus on the whole student with an emphasis on records, interviews, and the "scientific methods" the deans of men had long avoided.

However, despite Cowley's pronouncements and Bradshaw's urging, the deans of men were slow to change. They argued about "scientific nuts" and artificiality well into the 1940s. The attack by the Japanese on the U. S. Naval base at Pearl Harbor, Hawai'i ended the debate for a while. But the discussion and the debate about the future of the deans of men would continue when the war ended. F. F. Bradshaw, who as a young dean of students at the University of North Carolina had forecast the future of the work of deans of men and deans of students in 1921, might take some small satisfaction that he would finally see his early adoptions of change as the personnel movement began to at least supplement if not replace the older concept of the dean of men across the country.

When World War II broke out, the University of North Carolina, under Frank Graham as president, chose to create a War College on campus. The idea was to mobilize students and faculty alike in a patriotic response to the Japanese attack and the threat of the Axis powers, Germany and Japan, to the United States. Bradshaw was made Dean of the War College and a number of classes were quickly created in response. Under the direction of President Graham, the trustees pledged to improve aviation facilities for instruction at Raleigh and Chapel Hill, establish required military training for male undergraduates at Chapel Hill, and undertake a program of mandatory physical instruction until such time as military training could be established.[43] The War College was to be Francis Bradshaw's last significant act at his alma mater.

When the war ended, so did Bradshaw's long affiliation with the university and Chapel Hill, North Carolina. Bradshaw left the field of education in December, 1945 and moved to private practice as an industrial psychologist in New York City. He helped organize the firm Richardson, Bellows, Henry, and Company and was made president. He resigned as president in 1959 but remained active as Chairman of the Board and as a consultant. He died in California in 1959 at age 85 but was buried in North Carolina.[44]

Bradshaw was a significant character in the transition of the deans of men from reactive agents of the university to pro-active and engaged agents of change. Bradshaw might well have worked for or with Walter Dill Scott in Scott's consulting company created after World War I had his career in industrial psychology begun 25 years earlier. Clearly the two men saw the world in very similar ways. Both believed in the ability of vocational guidance and more specifically, industrial psychology and its application to channel human resources effectively and efficiently, whether it was on the college campus or in business, industry, or in Scott's case, the military. Bradshaw was a very ambitious man who was often of two minds; he was a faithful and conscientious dean of men/dean of students from 1920 well into the 1940s. At the same time, he saw the logic and usefulness of the personnel movement that became the student personnel movement on the college campus.

Bradshaw interacted in the NADM meetings with Thomas Arkle Clark, Stanley Coulter, and other early deans of men who often railed against the artificiality and "scientific nuts" who promoted the personnel movement. His loyalty to the organization and to the deans of men who attended the meetings was consistent over most of the years he was dean of students at North Carolina. As noted earlier, he served as an executive officer of the NADM (Secretary) for three years at the meetings in Chapel Hill, Minneapolis, and Atlanta.

At the same time, Bradshaw was also a clear proponent of the personnel movement on his own campus. Early in his career as Dean of Students, he made significant changes in the structure of the Dean of Students Office, embracing many of the new techniques and mechanisms of vocational guidance and student personnel that would slowly creep into the administration of other colleges and universities over the next 30 years. Bradshaw was clearly an "early adopter" of the student personnel movement, well ahead of the ACE-sponsored monograph, *The Student Personnel Point of View*, published in 1937, which many document as the real beginning of the move toward a modern student affairs era.[45]

Bradshaw was deeply involved in the personnel movement and working alongside Herbert Hawkes, L. B. Hopkins, E.K. Strong, W. W. Charters, Ben Wood, and C. R. Mann among others.

Though it is not clear from the minutes of the deans of men conferences who invited them, L.B. Hopkins, Ben Wood, and C. R. Mann all spoke at NADM meetings in the mid-1920s. The minutes of the 1926 NADM conference held at the University of Minnesota identifies that specific portions of the conference were held "in joint session with the Educational Personnel Workers." The session was titled, "Selection, Orientation, Educational Classification, and Guidance." The narrative recorded in the minutes of the meeting note that Dean C. R. Melcher of the University of Kentucky, who was presiding over the meeting as NADM president, welcomed the personnel workers to the deans of men conference. In his introduction, Melcher noted that this "was the first joint meeting of the two groups whose members were working toward similar objectives with dissimilar methods," a comment that served to highlight the division between the personnel workers and the deans.[46] As Bradshaw was one of two executive officers that year, serving as Secretary for the NADM, it seems too great a coincidence to ignore his influence in bringing the two groups in which he had invested so heavily together, even briefly.

Despite these brief intersections, the early deans of men remained committed to their own "dissimilar" methods for some time to come. In Appendix E published in the minutes from the 16th Annual Conference of the NADM, a compilation of the speaker's topics at the conferences from 1921 through 1933 is provided. At least 19 speakers gave formal papers on the "Functions, Qualities, and Definitions of the Dean of Men" over those years. Sub-topics related to the work of the dean of men ranged from salaries, budgets, and office staff to specific tasks of the dean, including fraternities, disciplines, student government and more, accounted for 120 papers. But under the topic, Personnel Work, only nine papers were given. One more listed under Vocational Guidance made the total an even 10. The conclusion can only be that the personnel movement was not a topic of great interest to the deans of men by the mid-1930s. However, it is quite clear that as a group, the deans of men were well aware of the increasing emphasis on the student personnel movement.

6

A Modern Dean: Fred Turner

Image 4 Fred Turner. Courtesy of the University of Illinois Archives.

Fred Turner was in his early twenties when he first began to work for Thomas Arkle Clark at the University of Illinois in 1922. Clark had developed a pattern of hiring mature, young men to work in the Dean's office under his direction. Calling themselves, "Tommy's boys," the young men arranged appointments, answered correspondence, filed reports and the voluminous paperwork that the Dean's office began to accumulate as the bureaucracy of the university, and in particular, the dean's office, expanded over time. Many of the young men stayed longer than they might have imagined at first.[1]

These staffing arrangements continued to evolve as the years and responsibilities rolled on and the Office of the Dean of Men expanded. By 1922, Clark was supported by Robert Gardner Tolman, who served as Dean for Freshmen. Turner became Assistant to the Dean. In 1923, James G. Thomas was added as Assistant Dean. In 1924, Turner ascended to become Assistant Dean and Thomas became Assistant Dean for Freshmen. By 1925, Roger Hopkins was added to the staff as an Adviser to Student Organizations and Activities.[2]

Clearly, there was a deep affection and appreciation among many of these young men for Dean Clark. Clark, for his part, treated the young men with respect, fondness, and genuine interest in their personal and academic well-being. In some respects, it would be easy to consider the young men in his office as Clark's surrogate sons, as Clark and his wife never had children of their own. From Clark's interest in fraternity life and affiliations, it is clear that he valued and sought out male comrades throughout his life. The young band of men who shared his office became a small fraternity of their own, complete with rituals, initiations, and long-standing commitment to each other and especially to "the Dean."

Fred Turner, a native of Hume, Illinois, a small community southeast of Champaign, Illinois, had no intention of becoming a dean like Thomas Arkle Clark. Turner had his sights set on a career in medicine. He first met Dean Clark when he, Turner, "came as a student [in Urbana] from Tuscola [IL]. I was seeking work. I got off at the Illinois Central Station but I had to walk to campus as there was a street car strike in progress. I had no trouble seeing Dean Clark. He was a small, rather dapper man who had a big office with a big desk. He was friendly and pleasant. He solved my problem [finding a job] quickly."[3]

During his time at Illinois, World War I began. When the United States entered the war, Turner enlisted in the Student Army Training Corps as a private instead of entering the military draft. His enlistment earned $30 a month but after the Armistice, the SATC ended. So Turner was once again in need of a job to pay for his schooling. His older brother had not been able to enlist in the SATC and instead, had found a job doing clerical work in Dean Clark's office for ten cents an hour. When Fred Turner returned to campus after Christmas vacation in 1922, he, too, began working in Clark's office for the same minimum wage.[4]

He earned his bachelors degree in chemistry at Illinois in 1922 and considered himself a pre-med student. However, Arthur Ray Warnock, Clark's assistant the year before, wrote to Clark indicating his intent to stay at Pennsylvania State College instead of returning to Illinois. So Clark asked Turner to delay his admission to medical school for one year and to be his assistant instead. Turner did so and also claimed that Clark encouraged him to go to graduate school. Turner subsequently earned a masters degree in Psychology in 1926 and completed a PhD in education in 1931.[5]

After working in the Dean of Men's Office full-time for two years, Turner was able to accumulate enough savings to marry his wife, Elizabeth (Weaver) in June 1924. He became Acting Dean in 1931 when Clark became ill that year. In 1932, after Clark's death from an intestinal cancer, Turner was named Dean of Men and succeeded his mentor after an apprenticeship of ten years in the Dean's Office.[6]

Fred Turner was more than a valued Assistant Dean to Clark. He became almost a member of the family. When Clark died in 1932, it was Turner who assisted Mrs. Clark with funeral arrangements, including the location of the gravesite for both Dean Clark and later, Mrs. Clark. Turner was involved in helping to manage the various honors and eulogies that both Illinois alumni and a national audience wanted to bestow on Dean Clark, and other issues related to his passing. Turner was also responsible for financial matters as executor of Clark's estate. He noted in an interview in 1967 that Clark had been "a sharp investor, especially in the last ten years of his life. He left a sizeable estate."[7]

When Thomas Arkle Clark died in 1931, a huge hole was ripped open in the culture of the University of Illinois. Clark had been in administrative roles on campus for over 30 years, serving as Dean of Men for 23 of them. Clark had, by his own definition, defined the work of the dean of men nationwide. Many other deans emulated Clark and saw him as the "dean of deans." It takes little imagination to consider what the loss of a "living legend" on his own campus at Illinois must have been like for students as well as faculty and staff.

But the steady and familiar hand of Fred Turner, Clark's hand-picked successor, alleviated the situation. A long-faced, serious-looking young man with round, wire-rimmed glasses stared out from pictures of him taken in the 1920s. Turner knew the lay of the land in the Dean's Office well. He had attended the national meetings of the deans of men along with Clark over the years and was familiar to a national audience as well as the faculty and staff at the University of Illinois. Initially, Turner continued to operate the office of the Dean of Men in the same manner as Clark but over time, he gradually reshaped the role of the dean of men in his own image.

By the time of his death, Clark had created an empire within the administration of the university. Over time, as Clark settled himself in the office from 1901 on, Clark had asked a succession of presidents for more money, more space, and a larger staff, to assist him with the increasing responsibilities of his office.

From 1901 to 1909, Clark was Dean of Undergraduates. In 1909, Clark's title changed to Dean of Men.[8] In 1912, Professor A. R. Seymour was appointed to be Adviser to Foreign Students and reported to Clark. In 1913, Arthur Ray Warnock joined the staff as Assistant Dean of Men. In 1914, the Dean of Men Office moved to the new Administration Building. By 1919, Warnock moved to Pennsylvania State College and was replaced by Horace B. Garman. Two men were assigned to serve as Assistant Deans for Foreign Students, Seymour in the 1st semester followed by David Carnahan in the 2nd semester.[9]

By the time Fred Turner was hired in 1922, Robert Tolman was the Assistant Dean for Freshmen, James G. Thomas was Assistant Dean, and Turner became Assistant to the Dean. In 1923, Robert Gardner Tolman became the Assistant Dean for Freshmen, James Thomas was Assistant Dean, and Turner was still Assistant to the Dean. In 1924, Turner became Assistant Dean, and Thomas was made Assistant Dean for Freshmen. In 1925, Roger Hopkins was added to the staff to work with Student Organizations and Activities.[10] By 1926, Thomas Clark reported that Illinois had a population of male students on campus of 7,500, equaling that of the University of Michigan. The next closest male enrollments were 7,000 at Pittsburgh and Minnesota followed by 5,000 at the University of Wisconsin.[11] These numbers were large and impressive. On many private campuses, such as Beloit, Oberlin, and Stanford, the total number of male students in 1926 was 350, 625, and 2,500 respectively.[12] So in terms of longevity in office and the number of male students, Thomas Arkle Clark was a significant figure among the deans of men. Because Turner worked for Clark and became, in essence, Clark's right-hand man, he could bask in the same light. The Office of Dean of Men at Illinois was a precedent setting and exemplary office for many of the other deans around the country.

Thomas Arkle Clark not only built a sizeable staff, he increased his responsibilities as Dean of Men at Illinois over time. His office was a bustling and impressive place. In 1928, Clark and his staff began to take photos of new students as they entered the Office of the Dean of Men (which all new male students had to do to register) to keep better track of the volume of students. Clark was also eager to know the names of the students or to refer to a face, not just a name, as a student was referred for advising, employment, and other concerns.[13] On a given day, the number of students who passed through Clark and Turner's office could be staggering.

At the Deans of Men's conference in 1936, Turner reported that the Deans Office at Illinois recorded 120,000 student contacts per year and was supported by an office budget of $34,000 ($441,159 in 2010 dollars).[14] A large portion of that budget went to pay for the 35 part-time clerks the office employed. The clerks were, for the most part, college men paid between 30 and 40 cents per hour for their attention to record-keeping such as absences from classes, a task which by itself accounted for some "one million and a half operations per year," according to Turner. Others tasks included recording missed class excuses, filing the excuses, filing attendance slips, visiting hospitals, and much, much more.[15]

Not only was the Dean's office busy, but the scope of responsibility for campus activities was impressive. Clark had assumed additional responsibilities for the foreign students on campus. He had added another staff member to oversee the many student organizations and clubs. He had also lobbied the administration to add an infirmary for students who became ill.

Clark had consistently encouraged the growth of the fraternity system. A staunch advocate of fraternities in general and a very active member of Alpha Tau Omega himself, Clark believed that fraternity life helped shape a young man both, for the responsibilities of campus life and after graduation. Fraternities had expectations for their members, rules to abide by, and they taught members to respect those in authority, a valuable intake process for new freshman at Illinois.

At the National Education Association meeting in 1910, Clark had expressed his belief that "... I have favored fraternities, and other social organizations... because I have found them of the greatest help to me in controlling and directing student activities."[16] Fred Turner was a member of Sigma Alpha Epsilon, a fact which no doubt further endeared him to Clark. Turner, too, eventually served as national president of his fraternity, although not until 1943. Clark's support of the fraternity system transferred easily to Turner when he became Dean, as he shared the same sentiments as his mentor toward fraternity life.

By 1930, near the end of Clark's career and his life, the number of fraternities at Illinois had reached 92 chapters. This growth of the system was no fluke; it had been a deliberate effort by Clark and Turner to expand the number of fraternity chapters as a means of managing, some might say "controlling," the large number of undergraduate students on campus.[17] Sororities grew as well, with 33 sororities on the Illinois campus. Nonetheless, a Greek system of 125 chapters by 1930 was impressive and very large by national standards.

Fred Turner began keeping a diary in 1930. His earliest thoughts were scattered and reflected some of the frustration he experienced in trying to keep the Dean of Men's office under control. Turner expressed the challenge of responding to the daily events in the Dean's Office while at the same time working on his courses and, most importantly, his dissertation. Often, he would conclude a diary entry for the day with comments on the end of his day—"to bed after 11p" or "to bed with the chickens."[18] In essence, by the time Fred Turner inherited the Office of Dean of Men from his mentor, Thomas Arkle Clark, in 1931, he inherited not only an administrative office but an administrative empire. Turner was not Dean Clark but he shared many of Clark's beliefs and attitudes. It is clear from Turner's comments in a 1967 interview with a reporter from the campus radio station, WILL, that he remained a staunch defender of Clark and his reputation, even 36 years after Clark's death.

In the radio interview, Turner acknowledged that "people either liked or disliked him. Those who disliked him did so because of something they had done and he (Clark) became involved in as chief disciplinary officer."[19] On the contrary, Turner argued, Dean Clark had been a "very kind man." Clark was "soft on money matters and aided students. He tried to become acquainted with them. He had an introduction slip before the students entered his office (prepared by his staff from university records.). He had memory systems and tricks to remember names and associations." Turner also claimed that "Dean Clark tried new things long before others, e.g. getting a university hospital" and regularly calling on students who were ill or having operations. Other developments that Turner credited Clark with included introducing "psychological tests for colleges (1912), a foreign students office (1910 or 1911) and the freshmen orientation program (1912)."[20]

Not surprisingly, as Clark's assistant and then his "right-hand man" and mentee, Turner kept many of the same functions after he became Dean. Other deans of men across the nation, many of whom knew Turner from the NADM meetings and from his association with Clark, sent him congratulatory notes after he was officially named Dean in 1932. Over time, he made changes and accommodations for the inevitable changes in campus life and university administration. Like his mentor, Turner also became a regular participant at the meetings of the National Association of Deans of Men.

Beginning in 1930, Fred Turner began to represent the University of Illinois at the NADM meetings as Assistant Dean of Men. At the 12th meeting held in Fayetteville, Arkansas, in 1930, Turner reported on the "Office Staff, Records, and Organization In the Dean of Men's Office" His presentation documented the vast array of tasks performed in the Dean's Office and supervised by Clark and his able assistants, including Turner.[21] While Dean Clark returned for the 1931 NADM meeting held in Gatlinburg, Tennesee, it would be the last meeting he would attend. From 1932 on, the University of Illinois was represented by Fred Turner.

At the 1934 NADM meeting, Turner used notes and correspondence between Clark and Scott Goodnight to fill in the "missing" minutes from the first two meetings of the Deans of Men meetings held in Madison, Wisconsin in 1919 and Champaign-Urbana, Illinois, in 1920. Much of the correspondence between Clark and Goodnight was readily available to Turner in Clark's papers at Illinois so it was simply a matter of organizing the material and presenting it to a very appreciative audience at the annual meeting. In closing the gap in terms of the missing correspondence and minutes of the meetings, Turner was able to pay further homage to the memory of Clark by documenting his mentor's involvement in the creation of the NADM from its inception. By 1936, Turner was again presenting at the annual meeting, this time as the chair of a new committee, a committee assigned to discuss the preparation necessary to become a dean of men. This topic was not new; it had come up at NADM meetings frequently over the 18 years the Association had been meeting. Turner's committee had taken on the unwieldy task of polling the membership of the dean's association and asking them to complete a lengthy survey. Turner and his committee had investigated the best means for a prospective dean of men to gain training in the field.

Turner's presentation takes up some 36 pages of the minutes from the 1936 Secretarial Notes of the NADM meeting held in Philadelphia. His committee, appointed in 1935, consisted of five deans—Turner,

James Armstrong of Northwestern, Harold Speight of Swarthmore, J. Jorgen Thompson of St. Olaf College in Northfield, Minnesota, and George Culver of Stanford. In addressing the topic, the preparation of deans of men, Turner first reminded the assembled deans of the legacy of the deans of men and the lengthy discussions that occurred in past years on the same question—what sort of preparation is necessary to be a dean? At the beginning of his lengthy presentation, Turner recounted, verbatim, comments from deans at previous meetings including Dean Stanley Coulter of Purdue in 1928 at the NADM meeting held in Colorado, who declared, "the Dean is a personality, not an officer." Robert Rienow of Iowa at the same meeting agreed and added that a dean must have "a great fondness for young people." In 1930 at the Fayetteville meeting, James Armstrong of Northwestern (but by 1936, re-located to Stanford) argued that "We should take the leadership in efforts now being made to offer academic courses pertaining to our work.[22]

At the 1931 meeting in Gatlinburg, Tennessee, Turner recounted, the preparation issue had been addressed by several deans. Among them was Robert Clothier of Pittsburgh, who stated that "I am afraid that I am not in sympathy with the idea of any fixed course in training for the position of dean of men.... The best preparation for the work is a broad general education, followed by years of experience in dealing with young people, and accompanied by a determination to take an optimistic view of the world and its problems."[23] Clothier's lengthy presentation concluded, "so if he gets the right start and has the patience, humor, courage, and sympathy already mentioned, what he needs is an open mind, a broad general education and a fund of experience. If he hasn't these qualifications, no amount of theoretical training in courses... will do him any good in my opinion. He is either cut out for the job or he isn't."[24]

Turner also recounted similar comments from Thomas Arkle Clark, Dean C.R. Melcher of Kentucky, Rienow of Iowa, and finally, Dean Francis Bradshaw of North Carolina. Of all the deans responding in 1931 to the issue of preparation for the work of dean of men, Turner noted, Bradshaw did argue for a specific set of courses. He suggested, "Study of Procedures, Study of Affiliated Specialties [to include] Mental Hygiene, Vocational Guidance, Educational Guidance, Testing, and Psychological Work, Philosophy and History of Education, and Educational Sociology." Bradshaw further suggested that there was much "in the sciences of psychology, sociology, and education... that deans could study with profit during leaves of absence and summer terms."[25] As noted earlier, Bradshaw had followed this course of action himself, by attending Columbia University through leaves and summers to complete his own doctoral degree in psychology.

Turner also cited his fellow committee member, James Armstrong, who, in 1931, had suggested that because there was an increased interest in training and preparation for deans, it might be good if the National Association [sic] took on some responsibility for a summer training program. He suggested Chicago as a natural site and offered a list of some 13 activities that might provide beneficial training to deans. Among the items were, "trips to Juvenile Courts, Psychopathic Hospitals, etc.; Contact with methods used by Northwestern's Personnel Department and Dean of Men's Office; visits to the art galleries of Chicago and other places with aesthetic interest; studying and working out articles on student behavior, the use of the interview."[26]

Finally, Turner also noted that at the 1932 conference held in Los Angeles, Dean Donfred Gardner of the University of Akron had presented the findings of a committee that had attempted to catalogue the functions of deans of men. Saddled with the unwieldy title, "Report on the National Survey of Functions of Student Administration for Men in Colleges and Universities of the United States," Gardner's committee had sent out a survey with 54 functions listed. The functions ranged from 'determine admissions" to "interview entering students for personal history records" to "administer student loans" and "penalize students for chapel or assembly absences." For each function, many of which were conducted by the Office of Dean of Men, the survey then attempted to determine what courses, if any, were appropriate to learn how to carry out the function. For example, "determine admissions" was linked to "Education" whereas "interview entering students for personal history records" was tied to "education and psychology." However, "penalize students for chapel or assembly absences" was considered to be a lesson learned outside the classroom as it was linked to "No specific courses," as were a variety of other functions. Overall, the final tally showed "No specific courses" linked to 15 functions, Education courses linked to 16 functions, Psychology to 10, followed by a variety of fields from business to accountancy to public health and religion.[27]

After this extensive introduction, Turner returned to his discussion of the role of deans of men. He noted that a number of changes in the functions of the dean of men had occurred over the seven years since the last such assessment had been completed. The most frequently reported changes had to do with the expanded responsibilities assigned to the dean's office, for example, counseling services, health service, employment supervision, centralizing student organizations, administering the National Youth Administration program.[28]

Turner presented the results of the 128 responses to a survey the committee had sent to 175 deans of men. The deans had been assigned a variety of titles. Although most were "deans of men," 11 had the title "dean of students" or "dean of student life." Of the deans who responded, 104 had been teaching in a college or university and 45 had been in college administration prior to becoming a dean. When Turner and his committee asked which courses had been most valuable to these men in preparation for their work as deans, psychology was listed first (62), education was second (45), liberal arts courses (31) and sociology (30) came in third and fourth, respectively.

Turner's fifth survey question asked, "Some outstanding deans have stated that a Dean of Men has inherent qualities in his personality which qualify him for his work, regardless of his preparation and departments of study. What are these qualities? What is your reaction to the statement that a Dean of Men is born and not made?" In response, some 25 agreed with the "born, not made" statement whereas 75 agreed with the qualification that training could be of definite service, 22 disagreed with the statement, and 6 did not respond.[29]

The survey next asked what inherent qualities a man should have to be a dean. The resulting list became extensive, running to two pages of individual attributes that Turner categorized loosely under "social traits" (e,g., congeniality, good mixer), "temperamental traits" (e.g., backbone, energy), and the like.

Question six asked the deans to cite general and specific courses that might be helpful to others in preparing for the dean of men's position. Under general courses were listed "broad liberal arts course (57)," "any academic subject as a major field (17)" and "graduate work essential (10)." Education (123) and psychology (114) were the top two specific courses recommended.[30]

When Turner and his committee asked the deans what practical training they thought would be helpful, the top responses were "apprenticeship to a Dean of Men (68)," "work with activities (43)," "administrative duties (30)," "counseling and interviewing (27)," "dormitory proctor (18)," and "business experience (16)." Other suggestions ranged from "Y.M.C.A. work" to "grade tests" to "speaking in public."[31]

Toward the end of his presentation, Turner (1936) presented a listing of "some of the courses now being taught" in the field, including the institutions where the courses were offered and the faculty who taught them. In addition to supplementing Turner's report, the list provided an interesting perspective on the status of professional preparation programs for deans of men and women in the mid-1930s.

Courses Taught Related to the Preparation of Deans of Men and Women in Selected Colleges and Universities, 1936[32]

COLUMBIA UNIVERSITY

1. Education s200D Demonstration of techniques in Guidance
   Professors Lloyd-Jones, Gates, Hollingsworth, Kitson, Lambert, McDowell, Pintner and Rowell, Doctors Anderson, Driscoll, Flemming, Hildreth and Gilpin
2. Education s200MG Orientation course in individual development and guidance Professors Kitson and Sturtevant, Doctor Flemming, and others
3. Education s200GA The teacher's part in individual development and guidance
   Professor Strang, Doctors R.N. Anderson and Driscoll
4. Education s238M Student Personnel Administration
   Professors Sturtevant, Lloyd-Jones, Strang, O'Rear, Linton, Hayes, Doctor Brown
5. Education s249V Vocational and educational guidance—Professor Kitson
6. Education s207M Psychological Testing—Professor Symonds and specialists
7. Education s237x Field work in student personnel administration—Professors Sturtevant, Strang, Lloyd-Jones and Doctor Brown
8. Education 249x Field work in guidance and personnel—Doctor Anderson
9. Education s250N Analysis of vocational activities—Professor Kitson
10. Education s249T Vocational testing—Professor Kitson and Doctor Anderson
11. Education s1490 Illustrative lessons in vocation and educational information—Doctor Lincoln
12. Education s249O Methods and content of the course in occupations—Doctor Lincoln
13. Education s349E Appraising the results of guidance—Professor Kitson, Doctors Anderson and Lincoln
14. Education s300GT Methods and techniques in guidance and personnel—Professor Strang and Doctor Anderson

15. Education s337Hn Special problems in student personnel administration—Professors Sturtevant, Hayes, O'Rear, Lloyd-Jones, DelManzo, Linton and Doctor Brown
16. Education s337Hn Special problems in student personnel administration—Professors Sturtevant, Hayes, O'Rear, Linton, Lloyd-Jones, and Doctor Brown
17. Education s337Hg Personnel records- Professor Lloyd-Jones
18. Education s337Ho Financial Aid to students—Professors Hayes, DelManzo
19. 160.9 Personnel Administration—Professor Reed
20. 160.18 Vocational Information, Guidance and Placement—Professor Reed
21. 260.13, 14 Course for Deans and Advisors of Men and Women—Dean Hagelthorne
22. 360.9, 10 Research in Personnel Problems—Professor Reed

Northwestern University

Educational conferences

Deans and Guidance Workers, July 22 and 23
School Administrators July 8, 9, and 10
Leisure time, July 27 and 28

Ohio State University

1. 750a Fundamentals of Guidance—Mr. Stone
2. 752 Vocational Studies—Mr. Smith
3. 754 Administration of Guidance Programs—Mr. Clifton

The Psychological Clinic open

Purdue University

S124 Secondary and Vocational School Administration—Dr. Cromer, Professor of Agricultural Education

Pennsylvania State College

Ed. 453 Educational and Vocational guidance in Junior and Senior High Schools—Miss Wyland

University of California, Berkeley

Conference for Junior College Principal, Deans, and Teachers, July 1,2, and 3

University of California, Los Angeles

1. S160 Vocational Education—Mr. Jackey
2. S161 Problems in Vocational Education
3. S164 Vocational guidance—Mr. Jackey
4. S169 Vocational Guidance for Women

University of Chicago

345. Educational and Vocational Guidance—Woellner
July 10, 11, and 12 Conference of administrative officers of junior colleges, colleges, and universities
July 15–19 Conference of Administrative officers of public and private schools

University of Iowa

No courses offered in Education
211S Psychology of Educational Personnel—Assistant Professor Jones

University of Michigan

1. E101 Vocational Guidance—Professor Myers
2. E102 The social and Economic Background of Vocational Education- Associate Professor Murtland
3. E107 The Technique of Securing and Using Vocational Information—Associate Professor Murtland
4. E201s Seminar in Vocational Education and Vocational Guidance—Professor Myers
5. 110 Vocational Psychology—Associate Professor Griffitts

University of Illinois

1. Educational Psychology—Professor Camerson, Assistant Professors Dolchand, Potthoff, Mr. Peters; (S, Dr. Gregg)

2. Mental Hygiene in the School-Professor Griffith (S, Assistant Professor Potthoff)
3. 18. Educational Measurements, I and II—Associate Professor Odell
4. 41. Principles of Vocational Education (same as Industrial Education 41)—Professor Mays
5. 43. Mental Tests—Assistant Professor Potthoff
6. 101. Philosophy of Education—Professor Cameron, Assistant Professor Browne
7. 121. Educational Measurements—Associate Professor Odell
8. 123. Educational Statistics—Associate Professor Odell
9. 125. Advanced Educational Psychology—Professor Cameron
10. 14. Social Psychology—Professor Young, Doctor McAllister
11. 21. Character and Personality—Doctor Sears
12. 23. Abnormal Psychology—Doctor Sears
13. 34. Individual Differences—Professor Woodrow
14. 2. Social Factors in Personality—Mr. Ahrens
15. 4. Social Control—Assistant Professor Albig

UNIVERSITY OF PENNSYLVANIA

The Psychological Laboratory and Clinic
Dean: Paul H. Musser, Ph.D.
Normal children and adults applying for educational and vocational guidance now constitute the larger part of the clientele of the clinic

UNIVERSITY OF SOUTHERN CALIFORNIA

144. Vocational Guidance

UNIVERSITY OF WISCONSIN

Conference for Deans and Advisers of Women

Among those institutions offering courses that could relate to the preparation of deans of men, Teachers College, Columbia was first in this area, both alphabetically and in course offerings, with 22 specific courses either directly related to the work of the dean or supplemental courses in education, personnel, guidance, and the like. Several schools, such as Northwestern, the University of Chicago, and the University of Wisconsin offered summer institutes. In many cases, the emphasis was on vocational guidance or training.

As of 1936, only Teachers College, Columbia University, under the influence of Sarah Sturtevant, Ruth Strang, and Esther Lloyd-Jones, had placed specific emphasis on courses with "student personnel" in the title. Perhaps the coursework emphasis preceded the Student Personnel Point of View (1937) by design or it may have been that Turner's record was incomplete. Nonetheless, the Teachers College faculty anticipated the coming trend in the dean's work to a student personnel focus with great accuracy.

Surprisingly, the University of Illinois was second in offerings only to Columbia, although the courses at Illinois do not appear to have been as specific to the preparation of deans or personnel workers as were the courses at Columbia. Although Turner may be allowed some latitude in representing his own institution, he appears to have used a less rigid interpretation of his own survey in reporting a large number (15) of education courses offered at Illinois as appropriate for the aspiring dean.[33] At the same time, Turner had completed his own doctoral work in psychology at Illinois only a few years earlier, so his personal knowledge of the curriculum at Illinois may have given him a distinct personal advantage in determining which courses might be relevant.

In the conclusion to his presentation, Turner reported that the general consensus, "up to the present time" among the deans of men had been against a fixed course of training for deans of men. Further, it seemed unlikely that all of the functions of a dean of men could be served by any existing courses although some, such as education and psychology courses, might supplement the dean's inherent qualities. In general, Turner's report and the NADM as an organization, seemed to conclude that despite their best efforts to plumb the depths of academic offerings and existing programs, the best deans were still "born, not made," an argument espoused for some time by several deans.[34]

In a discussion that followed Turner's presentation, Harold Speight, Dean of Men at Swarthmore and a member of Turner's committee, suggested

> ...there is no specific direction upon which we could expect general agreement in preparing men to serve as Deans and Advisers of Men. It seems to me that the greatest hope lies in the development of a profession, in its initial stages at any rate. My own interest lies in the suggestion that we should build up the profession through apprenticeships, and I wonder whether it might not be a good thing for us, as an association, to maintain, in the hands of the Secretary, a list of young

> men who are now in training, young assistants, with a specific report upon their preparation, their work when in college, what they are actually doing under the man whom they are now assisting.[35]

Speight's suggestion met with considerable discussion, most of it positive. Turner concurred and confirmed that his mentor, Thomas Arkle Clark, had taken the same approach himself. Turner further suggested that a second list of possible openings be kept by the association as well. Taking the other side of the issue, Dean Massey of the University of Tennessee countered,

> I am very thoroughly sold on the matter of training for deans of men, as I am also sold on the inherent qualities required in deans of men.... [However] I do not think that I shall ever give to the Association of Deans of men the statement that I need an assistant. I may sit down and write to some of these men and say that there is a possibility of a place here, in confidence, but I certainly wouldn't give it out where it would be broadcast because a man who has everything on paper to make a good Dean of Men has nothing at all in his clothes that makes him a good Dean of Men; he just doesn't fit.[36]

Another speaker, referring to Massey's comments but not identified, stated that he agreed with Dean Massey on the training issue, in that there was no one course or method that would make a man into a dean if he was not capable, but also concurred that "it might be quite valuable to a dean of men to acquaint himself with the instruments of personnel, if he realizes that this knowledge alone isn't going to solve all of the problems for him. It seems to me that we at least should have an acquaintanceship with these techniques or instruments, if some of these things are not to be taken out of our hands."[37]

Though the Turner committee acknowledged the increasing number of graduate level courses available, many of the deans present at the conference still argued that practical, on-the-job training for an aspiring dean of men was still the best course of action. Several of the younger deans active in the NADM, such as Turner or Francis Bradshaw, had earned a doctoral degree in psychology in the 1930s, whereas most other deans with advanced or even terminal degrees, held them in their field of teaching, such as English or German. As most deans had learned the professional practice of dean of men through experience and not from a graduate degree, it was more familiar and common to espouse the benefits of experience versus the artificiality of an academic program that could only theorize about student interactions and late night phone calls.

THE PERSONNEL ISSUE

Although the deans of men themselves may not have been severely affected by the Depression, many of their students had been. Enrollments declined nationally in the 1930s as a direct result of the hard times following the Depression.[38] Those students who could afford to attend college often depended on the dean of men for a job, either on campus or off, to supplement their meager allowances.As these responsibilities and demands increased, the deans added these areas to their job descriptions. Far from the organized, "scientific" approach of the "personnel officer," the deans of men made decisions based on intuition and instinct. During the Depression, their responsibilities increased to meet the need for additional financial assistance and the part-time employment of students. Understandably, the personnel approach was tolerated but not embraced.

But even those deans of men firmly against the "personnel" approach to student services and advisement, they were well aware of the primary tenets of the personnel movement. Various proponents of the personnel point of view had attended various NADM meetings since the 1920s, beginning with Ben Wood from Teachers College in 1929 and followed in 1931 by a true champion of the movement, L.B. Hopkins.

Hopkins had been on the staff of the Scott Company headed by Walter Dill Scott after World War I. When Scott made the decision to return to his alma mater as president, he hired Hopkins to be his Director of Personnel. Hopkins, Scott, and a graduate student, Esther Lloyd-Jones, developed the student personnel movement on the Northwestern campus. Hopkins authored a monograph on the personnel movement in 1926. By the 1931 NADM conference, he had left Northwestern to become president of Wabash College in Crawfordsville, IN.

When he addressed the deans of men in 1931, Hopkins spoke on the "Nature and Scope of Personnel Work" and noted that,

> ...there are some people to whom the words records and efficiency and systems are synonymous. There are many people who feel it is only a cold blooded individual who can have a personal interview and set down on a card facts gained in the interview....The fact is that records are essential, not only for educational research but also for the daily operation of the school or college. Records that are accumulated painstakingly, and recorded systematically, and interpreted intelligently will reflect not only the whole point of view of the educational scheme but also furnish the basis for detecting the existing weaknesses of the system and point the way for its improvement.[39]

As a new college president, Hopkins outlined his ambitious personal effort to interview all of the new students at Wabash in the Fall, all the seniors in the Spring, and the sophomores in the middle of the year. Following good personnel procedures, he planned to keep a summary record of each of his interviews for future reference. In closing, Hopkins urged the deans to consider adopting some, if not all, of the personnel practices which he himself used regularly.

At the same 1931 session, Dean Robert Clothier of the University of Pittsburgh addressed "The Relation of the Dean of Men to Personnel Work in the Larger University."[40] Clothier had also worked for the Scott Company along with Hopkins. He agreed with Hopkins on the need to understand and appreciate the individuality of each student. But Clothier went on to state, "I cannot escape the strengthening conviction that it is the dean of men's job to assume initiative for the personnel work of the university."

In concert with the personnel advocates, Clothier encouraged his peers to consider that the dean of men must be interested and involved in "the individual student's development to the limit of his individual capacity for growth...from the aspect of his whole personality."[41] Clothier noted some fourteen areas that should be considered as a part of such a plan of operation, ranging from admissions to extracurricular activities to academic performance to vocational planning and post-graduation placement. He also outlined the institution's responsibility to the dean and to the students on the campus, to be manifested in sufficient support of staff, equipment, and time to collect the necessary information and research.[42]

Despite his very contemporary view of the dean of men's role as a practitioner of personnel methods, Clothier also reflected this thought,

> I am afraid that I am not in sympathy with the idea of any fixed course of training for the position of dean of men. I do not believe that it is possible to prepare oneself for this work by taking certain prearranged courses of study. It is not like preparing for the professions of law, medicine or engineering. To a very considerable degree the best and most successful deans of men are born and not made....I agree that there are some technical subjects such as mental hygiene and vocational guidance which require special training, but it is my theory that these areas should be handled by specialists who may or may not be a part of the staff of the office of dean of men, leaving the dean himself free to lay out the general policies of the office and to advise and consult with the many students who want to talk over their problems and difficulties with an older person, who will listen to their stories with a sympathetic understanding.[43]

Professor Lorin Thompson, Jr. of Ohio Wesleyan spoke after Clothier. He addressed the "Relation of the Personnel Officer to the Dean of Men in the Small College." Professor Thompson's primary point was that the personnel function and the personnel officer was a record keeper and an "expert technical adviser." The dean of men, on the other hand, was an officer of the institution and responsible for policy and direct interactions with student.[44] Following Thompson's paper was a general discussion by the deans. Many of the responses argued that the personnel officers were attempting to take away many of the duties of the deans of men and leave in their wake only the disciplinary functions. Many of the deans were opposed to this trend and said so. From the brief example cited below, it is fairly clear that the need for disciplinary action could consume a fair amount of time for both deans of men and deans of women.

Discipline of Women and Male Students, 1928 (probation/suspended)

	Women	Men
Grades	1802/861	7309/3441
Cheating	28/33	78/136
Stealing	3/7	13/56
Drinking	12/14	98/109
Automobiles	8/4	3/32
Sex	3/27	0/23
Debts	1/3	4/29
Gambling	not reported	9/17
Other	72/36	175/171
Total Enrolled	43,592*	93,223**
Total Disciplined	2,914	11,683
% of total disciplined	7%	12.5%

* Figures are for women enrolled at 31 reporting institutions.
** Figures are for men enrolled at 40 institutions.[45]

Clark had cautioned all deans, and no doubt Turner in particular, to not let themselves be turned into "scientific nuts" and ignore the real role of the dean of men, which from Clark's perspective was to work with and on behalf of the (male) students on campus. The rift between the "personnel movement" and the deans of men was polite but challenging for all concerned. To concede that training might enhance one's abilities seemed to confuse the issue of who a dean of men should be and how he was to perform. There was some legitimate concern on the part of the deans that they would be responsible only for discipline, the least attractive part of their jobs, but there was also the belief that the "personnel movement" and its functions would be subsumed under the Dean of Men on most campuses.

As a young man in his early thirties, Turner had already earned his Ph.D. in education at Ilinois. In fact, in his recollections of Clark, he insisted that Clark had encouraged him as well as other young men working in the Dean's Office to pursue graduate degrees. With Northwestern University in Chicago (where Walter Dill Scott implemented the student personnel effort along with L.B. Hopkins and Esther Lloyd-Jones in 1925) only 140 miles or a few hours away by car or train, Turner was very conscious of the personnel movement. Though Turner owed a great deal to his mentor, he had to acknowledge the oncoming train that was the student personnel movement.

Turner's efforts as chairman of the Committee on the Preparation of Deans of Men was, at the least, a conciliatory gesture to bridge the two camps, the personnel movement and the deans of men. Turner could see from his committee's work that there were many courses offered at Columbia. Even his own campus, the University of Illinois, reflected the "science" of the personnel movement through a large number of courses. Turner saw that perhaps the best option was moving the deans of men into a supervisory role over the personnel activities on campus. Given the breadth of the Office of Dean of Men at Illinois, it made great sense to add the personnel functions to those duties already assumed by the Dean.

The murmur of change became a crescendo at the NADM annual conference held at the University of Texas in 1937, when W.H. Cowley, professor of psychology and director of the Bureau of Educational Research at Ohio State University, presented his paper titled, "The Disappearing Dean of Men."[46] As might be expected, the title of Cowley's address, if not the content, was still reverberating throughout the NADM collective memory some thirteen years later in 1950.

Cowley's thesis was simple and straightforward. He claimed that as the need for student personnel services expanded within higher education, the office of the dean of men would cease to exist. Cowley cited the fact that "deans for student relations" and "instructional" or academic deans had "grown out of the same tree" under President Eliot's administration at Harvard in the late 1800s and both had changed significantly over time. Cowley anticipated the splintering of some student personnel functions among the deans of men and similar areas as an obstacle to effective work with students.[47]

To make his case clear to the deans, Cowley used the example of the admissions director who collected valuable information about an entering student upon admission via the college's application form.

This application form "...was replete with information of high value to the members of the personnel staff other than the admissions officer."[48] However, the form and its information probably remained in that individual admissions office, forcing others, if in fact they wished, to collect it all over again, wasting of their time as well as the student's.

In essence, Cowley noted,

> ...all student personnel services are but different types of the same sort of activity...a basic unity runs throughout them all. If the assumption is sound, then it follows, it seems to me, that somehow they should all be made to work together in unison, that they should all move forward in step, that, in brief, they should all be coordinated.[49]

Toward the end of this landmark presentation, Cowley listed the 22 functions necessary in an ideal personnel program. He suggested ways in which the coordinated delivery of such functions might be accomplished though four critical phases of a student personnel program: (1) the analysis of the functional list, with appropriate additions or deletions; (2) the assignment of an expert, such as one individual who would coordinate placement, each function or appropriate grouping of functions; (3) a review of gaps as well as overlaps in the functional areas; and (4) the centralization of the entire coordinated program under a director or coordinator.

Cowley closed by describing the "three roads" toward a coordinated program in student personnel that would be available to institutions and the likely place of the dean of men in such a re–organization. The first "road" was that the current dean of men would become the coordinator. Cowley cited two institutions: the University of California, Berkeley, where these changes had already occurred, and the site where the NADM meeting was taking place, the University of Texas in Austin, Texas, where change was about to happen.[50]

The second "road" was making the dean of men subordinate to a coordinator of personnel services. Cowley noted that this had been done recently at the Universities of Oregon and West Virginia. The third option or "road" was to do away with the positions of both dean of men and women entirely, as had been done at Earlham, William and Mary, Iowa State, and Northwestern.[51]

In closing, Cowley acknowledged that the deans of men might hope that a "fourth road," one in which things would remain much as they were, was possible. However, he insisted, this option was not realistic. The turn toward student personnel services in higher education was inevitable and universal. Cowley warned that though a clear majority of the deans of men might prefer the "first road" he had described, one in which the existing dean of men was named as the new coordinator or director of a revamped organization, much would depend on the incumbent. Each dean would have to be evaluated on his own merits. A successful transition would depend on the dean of men's "training, his temperament, his intellectual range, his ability as an executive, and . . . his spirit."[52]

Of the many questions Cowley fielded at the close of his 1937 presentation, one was particularly telling. A dean in the audience of men asked whether or not the dean of women would also be "subordinate" and, testing the waters, would it be possible for a woman to be promoted to the top spot? Cowley's response was simple and straightforward: "It seems to me there isn't any reason deans of women shouldn't go up if they are equal to it. I think we can say that the deans of women are in exactly the same position as deans of men."[53]

In retrospect, W.H. Cowley was iterating, with some courage given his audience, the primary tenets of the "Student Personnel Point of View" published by the American Council on Education later that same year. In fact, the subcommittee on Student Personnel would meet only two weeks later to endorse their own report and the monograph that followed. Cowley's presentation was not coincidence, as Cowley was an active member of the American Council on Education (ACE) and the chair of the sub-committee that wrote the monograph, "The Student Personnel Point of View."

The members of that committee included some names familiar to many of the deans of men, including F. F. Bradshaw, D. H. Gardner, L. B. Hopkins, Herbert Hawkes, and possibly Esther Lloyd-Jones. Thyrsa Amos represented the National Association of Deans of Women. George Zook, who presided over the ACE was well known as well. Of the total sub-committee of 19 people, at least three were or had been Deans of Men—Bradshaw, Gardner, and Hawkes. Dean Hawkes of Columbia never made it to a meeting of the NADM, although he was expected at least once, but Bradshaw and Gardner were regulars. Bradshaw served as Secretary for three years and Gardner held the office of Secretary for six years and became President in 1939.[54]

There were numerous responses to Cowley's "Disappearing Deans of Men" speech as might be expected. But in fact, the overall response was measured and muted. J.F. Findlay, Dean of Men at the University of Oklahoma, presented his paper on the "Origin and Development of the Dean of Men." Findlay would complete his doctoral work at

New York University and it is likely that his paper was a portion of a larger study, such as a dissertation.[55] Findlay's work cast a wide net, de-mythologizing some historical notes regarding the creation of the dean of men's office. Challengers to Clark's claim to be the first dean of men were raised. A gentleman at the University of Oregon, John Staub, had been Dean of Literature, Arts and Sciences since 1899, performing many duties, including some of those that would normally fall to the Dean of Men. But no official act to create a Dean of Men was taken at Oregon until 1920, according to several correspondents from that institution.[56] Other candidates included Deans of Colleges at Oberlin, Massachusetts Institute of Technology, and Beloit College (WI). Though the conclusion was that Clark probably retained the title of "first official dean of men," it was clear that other deans were probably doing the same tasks without the title.[57]

Findlay chronicled the background development of the Office of Dean of Men on a national scale, citing correspondence he had solicited from a variety of college presidents, a few deans of women (at least one of whom, Irma Voight of Ohio Weslyan, claimed that deans of men were created in response to the success of deans of women), a few association heads including the president of the American Association of Colleges, and the Carnegie Foundation, and several deans of men. Findlay also determined from correspondence with university representatives in Paris, Geneva, London, Florence, Bern, Berlin, Vienna, and Zurich that the Dean of Men was an American invention. Professor Paul Dengler of Vienna surmised that the reason for this was that in Europe, "our general education finishes at the secondary level."[58]

In conducting his study, Findlay had surveyed some 90 Deans of Men and gathered their responses to a range of questions. He asked about changes in staff and physical equipment in the dean's office; changes in title; and more recently, the dissolution of the dean's position. Several institutions confirmed that, in fact, that they had terminated the office of dean of men in favor of other positions. The institutions included Earlham College, a Quaker college in Indiana, where an Assistant to the President took over the duties. William and Mary established a Dean for Freshmen. West Virginia University [sic] made the Dean of Men into the Secretary for Loans, Placement, and Guidance. From Northwestern, Walter Dill Scott indicated that along with the Dean of Men, "the personnel service of the University was developed independently under a slightly different philosophy. As the work of these two offices developed, it became apparent that the separate organization of the two, both dealing with undergraduate affairs, was a mistake."[59] Scott indicated that "this last spring" which would have been 1936, both the Dean of Men and Dean of Women ceased to exist. Similar responses were received from the University of California, Berkeley and the University of Minnesota.

At the conclusion of Findlay's paper, Fred Turner addressed the assembled deans with his response to both Cowley and Findlay's paper on the origins of the deans of men. He stated that,

> I think that Dr. Cowley and Dean Findlay might have mentioned the fact that there is nothing especially disturbing about this whole situation. It is a point in evolution. Individuals on the faculty staff who showed a particular aptitude or interest in the student did the work [of the dean of men] at the outset probably unofficially. Student came to him because he was a fine fellow. Then came the time when the administration and the president found it was too much for them and they began to spot a man chiefly on personal qualifications who could help out. I can't see anything to be especially concerned about. It's too big [the job of dean] for one man to handle. It may be a group of officers under one man, but it's simply the next step when you get down to it.[60]

Turner's calm and circumspect demeanour about the "disappearing dean of men" was born of both his own personal experience as a member of Dean Clark's office at an early age and from his recent professional experiences in talking to and working with professionals from other organizations and associations. In 1936, Turner had presented an early version of his paper on the preparation of deans of men to the National Association of Deans of Women. He then shared a more complete copy of the same paper with the deans of men. Turner would soon serve as one of several men representing the National Association of Deans of Men (NADM) during three meetings on the "Coordination of Personnel Associations" between 1938 and 1940.

A loose confederation, the several associations agreed to meet together, to produce a brochure, and to begin to coordinate matters of mutual interest and importance. This coordinated effort was to be accomplished through a standing committee representing deans of men, deans of women, and the personnel associations, all the while maintaining the autonomy of the original groups. Another dean of men, F.F. Bradshaw, was already involved through his leadership in the National Association of Appointment Secretaries (NAAS), which later changed its name to NAPPO to reflect the new emphasis on "personnel" and even later to the American College Personnel

Association (ACPA). Bradshaw eventually became one of the founders of the confederation known as the American Personnel and Guidance Association, which grew out of the meetings of the personnel associations noted earlier.

In short, Turner could see the direction in which the personnel officers, deans of men, and deans of women were heading. While the deans of men might not disappear, as Cowley had suggested, they would, in fact, be transformed into a broader realm than they had occupied when Dean Clark became the first dean of men "officially" in 1909. Turner, no doubt, could hear Clark's words becoming estranged from the real work of the dean of men, regular and daily contact with students echoing in his ear, Clark had told him, "Don't let them make administrative officers out of you and put you on so many committees that you will not have time to see a student...."[61]

At the 1939 meeting of the deans of men, a change was clearly underway. Fred Turner had chaired another committee, this time charged with replicating the survey on functions of the dean of men, first completed by D.H. Gardner in 1932.[62] In presenting the results of the survey, Turner noted that a number of changes in the functions of the dean of men had occurred over the seven years since the last such assessment had been completed. The most frequently reported changes had to do with the expanded responsibilities assigned to the dean's office, for example, counseling services, health service, employment supervision, centralizing student organizations, administering the N.Y.A. program. By now, the ACE monograph, the Student Personnel Point of View, had been in existence for two years.

F. F. Bradshaw of the University of North Carolina, acknowledging that he had not attended an NADM meeting since the 1931 Gatlinburg, Tennessee, meeting, thought he sensed a trend his peers may have overlooked.

> I cannot help but feel the problem we are facing in deciding whether or not we are willing to be a staff office or a general executive office. I am convinced that the personnel point of view in the general personnel program is here to stay and expand for a while, and it seems to me that in that situation, the deans of men may either be specialists in discipline, social education or whatever you may choose to call it, or the causes of a more general type of training and personality may develop him into a general executive in which case, in the coordinated set-up, it seems to me that he might easily move in the direction of the central personnel coordination. But, he would have to make his viewpoint and training as broad as theirs is, in expecting to coordinate....If we go in that general direction, it is essential to associate

> more with other people in personnel guidance fields.... If we are to coordinate a general feeling, we must have more or less continued association with members of that same feeling.... I think we should consider a meeting at least every other year with the personnel and guidance associations.[63]

Some eight years earlier, Bradshaw had urged his fellow deans to consider a moderate program for training future deans of men, but others had countered with the deans of men are "born, not made" argument. Bradshaw's absence from NADM sessions had been spent in working closely with the American College Placement Association and the emerging Council of Personnel and Guidance Associations (CPGA) (later Association of Personnel and Guidance Associations (APGA)) board.

Still, not all the deans were convinced that cooperation was in their best interest.

At the 1939 meeting, D.H. Gardner (serving as the President of the Association for the year) reported on a meeting held in 1938 in his home community of Akron, Ohio. A group of sixteen people met at the invitation of the NADM executive committee to discuss their mutual concerns and how to cooperate. Eight deans of men (Deans Bursley of Michigan, Findlay of Oklahoma, Gardner Goodnight of Wisconsin, Stephens of Washington University, Lancaster of Sweet Briar, Lobdell of M.I.T, and Turner of Illinois) met with two deans of women from the NADW, and representatives of college union managers, admissions officers, and E.G. Williamson representing the American College Personnel Association. Their focus was on two issues: (1) how can individuals from these organizations cooperate on their own campuses and (2) how can cooperation be effected on a national scale.[64] So with Gardner's leadership and the cooperation of other executive officers, including Turner, the NADM was making an effort to take a leadership role with other student-oriented groups. By taking on the issue and seeking cooperation, the deans of men were putting themselves in a position to lead, rather than follow, the trend toward consolidation of the dean's role and that of the personnel movement.

Fred Turner was of a like mind with Gardner. He could see that the move toward the personnel movement was quickly being adopted on many campuses. He took Cowley's suggestion to heart as well—either adapt to the new directions of the student personnel movement and become a leader on campus and beyond or be left behind. However, Turner was loyal to the rudimentary training and beliefs of his mentor, Thomas Clark. Turner had been a confidante of Dean Clark as well as Mrs. Clark. His relationship with the two of them extended far beyond the work of the Dean of Men and it was a significant part of Turner's early career and college life. Turner's commitment to Clark and his work and later, to his memory, knew no boundaries.

It should be noted that Thomas Arkle Clark had a significant impact on many other men as well, not just Fred Turner. At his retirement and later, his funeral, testimonials to Clark and his compassionate work on behalf of students were overwhelming. Not only did former students offer their appreciations but many professional associates of Clark offered their strong endorsement of Clark as dean and as a man. So it is not surprising that Clark left a strong and lasting impression on Fred Turner.

To his credit, Turner did find his own way. Within a decade, Turner became his own man. During the transitions of the 1930s, Turner emerged from the shadow of Dean Clark and his multifaceted, occasionally notorious machinations and controls over students to become a leader among the many deans who were in the National Association of Deans of Men. Turner accepted the change toward the personnel movement and worked diligently with various other groups to effect a positive change both nationally and on a campus-by-campus basis. Though Turner may not have embraced the personnel movement to the same degree as his colleague from North Carolina, Francis Bradshaw, he was able to see the future and break away from doing his work in the shadow of Dean Clark.

Fred Turner remained in office at the University of Illinois for a long time. In fact, he never left. He served as Dean of Men from 1932 to 1943. Following the new conventions of the time, he was named Dean of Students in 1943 and held that office until 1966. Turner served as the Secretary of the NADAM, and therefore as a member of the Executive Committee, from 1938 to 1949. He eventually served as President of NASPA in 1958–1959. As noted earlier, he maintained a strong interest in fraternity affairs. He served on the Supreme Council of Sigma Alpha Epsilon from 1937 to 1943 prior to his term as national President of the fraternity from 1943 to 1945. He was Editor of the Interfraternity Research and Advisory Council (IRAC) *Bulletin* from 1953 to 1966 and a columnist for *Banta's Greek Exchange* from 1953 to 1968. Turner was on the National Interfraternity Coference Executive Committee from 1961 to 1967 and then served as NIC President from 1967 to 1968. Fred Turner died on September 6, 1975 at the age of 75.[65]

7

A Brief Treatise on the Deans of Women

No scholarly work on the role and activities of the deans of men can or should go too long without some mention of the deans of women. As at least one dean of women, Irma Voight of Ohio Wesleyan noted, the deans of men were probably created out of appreciation for the work of the deans of women.[1] In contrast to the deans of men, there have been a number of very good scholarly treatments of the deans of women. Two of them are books, Jana Nidifer's *Pioneering Deans of Women: More than Wise and Pious Matrons* published in 2000 and Carolyn Bashaw's *Stalwart Women: A Historical Analysis of Deans of Women in the South* published in 1999.[2] Three contemporary dissertations include Kathryn Nemeth Tuttle's What Became of the Deans of Women: Changing Roles for Women Administrators in American Higher Education from 1996; Lynn Gangone's Navigating Turbulence: A Case Study of a Voluntary Higher Education Association, and Schwartz's The Feminization of A Profession: Student Affairs Work in American Higher Education, 1890–1945.[3] There have been several articles in the recent higher education literature on deans of women as well. So the overview that follows is meant to provide only a brief, not exhaustive, perspective on the deans of women in comparison to the deans of men.

In point of fact, as Irma Voight noted, the deans of women did precede the deans of men historically and many were on campuses prior to the appointment of a dean of men. Though there were matrons and preceptresses on campuses in the late 1800s, the consensus holds that the first dean of women was typically considered to have been Alice Freeman Palmer, who served in the position of Dean of Women at the University of Chicago. She was appointed by William Rainey Harper and intentionally asked to be in office before the university opened its doors in 1892.[4]

Palmer brought her good friend and colleague, Marion Talbot, with her from Boston. Talbot originally served as Assistant Dean of Women. Eventually, she was promoted to Dean when Palmer left Chicago in 1895.[5] Other deans of women were appointed in the same time frame, 1890–1910, primarily to help the predominantly male-oriented colleges and universities of the time respond to coeducation. Palmer and Talbot were to represent the women students on campus and to be responsible for their welfare. The two deans were given the Beatrice Hotel, built for the World's Fair a few years earlier, which was converted into a residence hall for Chicago's female students until a proper women's hall could be built.[6]

In 1901 and 1903, Marion Talbot organized meetings with other deans of women from around the country. The intent of these early meetings was to share their experiences, discuss matters of mutual concern (and frustration), and plan for the future. These meetings became the first stirrings of a formal organization to represent the deans of women on a larger scale.[7]Alice Freeman and Marion Talbot had first met and become friends at early meetings of the Association of Collegiate Alumnae (ACA), a loose confederation of college-educated women organized by Alice Freeman's mother in Boston. The primary goal of the ACA was to encourage and promote the college education of promising young women. In the 1880s, women in Boston were not admitted to the Boston Classic Latin School, the acknowledged stepping stone to Harvard. In addition, Harvard president Charles Eliot, had made his opposition to admitting women to Harvard quite clear. However, he did encourage the creation of a co-ordinate college that could be a partner to Harvard, for the education of women who desired to go to college. Alice Freeman had earned her degree in history at the University of Michigan but found herself in Boston often as she had taken a position at Wellesley College and would eventually be named president of the College.[8]

As charter members of the ACA, Freeman and Talbot had developed a strong friendship. When Freeman left the presidency of Wellesley to marry Harvard philosopher, George Herbert Palmer, she also ceased much of her active, academic life. However, Harper was determined to have a nationally known woman has his dean for women, so he lobbied Alice (Freeman) Palmer exhaustively until she agreed to be in Chicago for a portion of the year and return to Boston frequently. She also talked Marion Talbot into joining her. So it was that the two women moved from genteel Boston to the frontier town of Chicago at the end of the nineteenth century to work for Harper at the newly created University of Chicago.[9]

Women were so eager to attend the new University of Chicago that the numbers quickly exceeded the expected enrollments. Women were academically inquisitive and quite competitive. Over time, they proved to be academically superior as well and threatened both the male students and faculty to the point that the men of Chicago petitioned Harper to separate Chicago into separate colleges, one for women and one for men, to keep the women from dominating the academic honors. Harper eventually relented but the "experiment" of separation only lasted a few years. As many frugal Midwesterners had determined long before Harper arrived, running two separate colleges based on gender was too expensive and unwieldy.[10]

As Marion Talbot demonstrated, the deans of women were eager to be successful and to organize themselves professionally. Much like the women students they supervised, many of the deans of women had been academically superior students. Most, if not all, deans held faculty rank and became deans of women out of administrative necessity. Beleaguered presidents and boards of trustees quickly conferred additional salary and other considerations on female faculty who would consent to also serve as deans of women. But beyond the appointment as dean, most of the women were not given elaborate definitions of their work. For that, they turned to each other. Meetings were often informal and many of the early meetings were held in conjunction with the annual meetings of the Association of Collegiate Alumnae (ACA). Many of the deans were members of the ACA or soon joined and while attending ACA meetings would meet in special session with other deans. Several enterprising deans even arranged to ride the train to the ACA meetings together to spend even more time discussing their mutual concerns.[11]

The ACA from its founding had sponsored research projects and activities designed to support the mission of the organization—to emphasize the value and potential of women in college and in the professions open to college graduates. Through membership dues and other means of support, the ACA sponsored needed research to rebut the notion that women were "less than" their male counterparts in many fields and occupations. As the deans of women developed their own organization, the inherent need for research on and about the work of deans of women and their impact on students seemed like a natural development.

By 1916, enough momentum had been created for a formal organization to be developed, and the National Association of Deans of Women was created. The first president was Kathryn Sisson McLean (later Phillips), a dean of women at Chadron State College in Nebraska.[12] Around this same time, many newly appointed deans of women had been attending summer school at Teachers College in Columbia University in New York. Many new deans were eager to take courses that might help them manage their new responsibilities as deans and lobbied the Teachers College faculty for such courses. In response, new courses were developed and a special focus on deans of women became a part of the curriculum. Over time, the new area drew a strong faculty, including Sarah Sturtevant, Ruth Strang, and later Esther Lloyd-Jones to Teachers College. The core faculty of Sturtevant, Strang, and Lloyd-Jones taught the deans not only to appreciate academic coursework related to their roles but to value research and publication that emphasized the value of women in leadership positions. So the nexus of graduate study and research between the ACA and the graduate courses at Teachers College was created and became a central part of the new NADW. [13]

The deans of women generated a significant amount of literature related to their work in a relatively short period of time. The first significant work was a book aptly titled *The Dean of Women*, published in 1915.[14] Written by Lois Kimball Mathews, dean of women and professor of history at the University of Wisconsin, the book detailed the work of the dean of women from individual counseling to classroom teaching and beyond. Mathews made her strong convictions about the value of women's work as deans and academicians clear in her book and it set the stage for the deans of women who would follow her. Prior to Mathews, Marion Talbot had produced *The Education of Women* in 1910.[15] Although her book did not address the work of the dean of women specifically, it did lay out a strong case for the education of women and was clearly a refutation of the efforts at the University of Chicago to minimize the role of women on campus. In 1936, Talbot and Mathews (by then Rosenberry) would co-author *The History of the American Association of University Women*, the latter-day name of the Association of Collegiate Alumnae.[16] The following year, 1929, *Deans and Advisers of Women and Girls* written by Anna Eloise Pierce, Dean of Women at New York State College for Teachers, appeared.[17] In this huge volume, some 600 pages long, Pierce attempted to cover all the duties of the dean of women and offered charts, graphs, recipes for the dining hall, and more. Despite the intimidating size, Pierce's book found an audience, as a second edition was produced the next year.

Soon the graduate program at Teachers College, Columbia, began producing its well-known Contributions to Educations Series and many female graduate students, under the tutelage of Sturtevant and Strang, began to produce monographs in the series. One of the first was by Jane Jones (1928) titled, *A Personnel Study of Women Deans in Colleges and Universities.* Her work was soon followed by others in the series. From the late 1920s through the 1930s and into the early 1940s, landmark studies, which documented the work of the dean of women, were published by Teachers College. Jones' 1928 work on deans in colleges and universities was equaled in the same year by Sturtevant and Strang's study of deans of women in teachers colleges and normal schools.[18] A subsequent study, published in 1929, also authored by Sturtevant and Strang, examined deans of girls in high schools.[19] These studies were for the most part, complex surveys of deans of women in like institutions, such as teachers colleges and normal schools. The data compiled was of great value in quantifying the experience of the emerging profession and setting baselines from which the common denominators of "deaning" [sic], such as salary ranges, specific duties assigned to the office of the dean, educational backgrounds, and extent of professional training, could be extracted. Much of this information was of value to the NADW and even more so to graduate programs, such as the pace-setting department of Deans of Women and Advisers to Girls which had developed at Teachers College.[20]

The National Association of Deans of Women had published a Yearbook, which also served as its professional journal, since 1927. The Yearbook of the National Association of Deans of Women carried papers, research articles, and addresses specific to the work of deans of women.[21] As chair of the Research Committee for the NADW as well as a faculty member at Teachers College, Ruth Strang extended her influence on the emerging profession of deans of women beyond the classroom. Strang and her colleague, Sarah Sturtevant, provided a great deal of the guidance and mentoring for female graduate students entering the profession. As chair of the Research Committee in the early 1930s, Strang was able to anchor the NADW to contemporary research in both secondary and higher education which could, and in her opinion should, influence the work of the dean. Doyle, in her study of the NADW Yearbook, noted that, despite her brief tenure in the chairmanship of the Research Committee, Strang's reports covered 66 pages, almost half the total pages committed to the Research Committee in all twelve Yearbooks published as of 1936.

In her preface to the report of the Research Committee at the 1934 NADW conference, Strang noted the position of dean of women has both artistic and scientific aspects:

> The artistic side is represented in the inspirational and philosophical articles; the scientific aspect in the description survey and experimental study of plans and procedures of work with individuals. It has been the self-imposed task of the research committee to summarize the investigations relating to the work of dean of women, and thus to make easily available annually the more or less scientific body of professional subject-matter published during the year.[22]

Some evidence of Ruth Strang's self-confessed Puritan "determination and work ethic" is apparent in her statement.[23] Following her preface was the actual research report, divided into two sections—one on secondary education and one on higher education—each supplemented by extensive bibliographies. In such reports, the Research Committee typically provided the NADW membership with brief reviews of current research in specific categories: "Admission and Orientation of Students," "Personnel Workers," "Personality, Attitudes and Achievement of Students," "Educational and Vocational Guidance."

In the final paragraph of the higher education section of the report for 1934, Strang editorialized,

> What lines and types of research are needed? Articles on guidance in educational magazines during the past five years have been predominantly descriptions of guidance programs and practices. Only 140 of the 461 articles analyzed involved some systematic investigation. Professors and directors of guidance emphasize the need for measures of the effects of guidance services and opportunities. Well-planned programs of guidance should be set up, groups of students followed through these programs, complete records kept at each step and the results carefully measured.[24]

The Research Committee's bibliography for the Higher Education section in 1934 alone included some 115 articles. Under Ruth Strang's leadership, comparable reports and bibliographies were prepared by the Research Committees and published in the Yearbooks in 1935, 1936, and 1937. Soon enough, the Yearbooks of the NADW began to be known for their inclusion of cutting-edge research on issues related to college, college students, and the scientific application of new concepts, including the personnel movement, to colleges and to students.

A new research approach to the role of the dean of women was implemented by Eunice Mae Acheson who, in 1932, published her study, *The Effective Dean of Women.*[25] Like Jane Jones, Acheson had conducted her research at Teachers College under the mentorship of Ruth Strang and Sarah Sturtevant. In her case however, Acheson attempted to determine the "effectiveness" of 50 deans of women. She developed a survey that she distributed to the 50 deans, and also to selected students, and the presidents of each dean's institution. Acheson had also collected vocational and personal histories on each of the 50 deans in her final sample.[26]

As a result of her research, Acheson determined that certain significant factors contributed to the "successful" dean of women, as judged both by presidents, students, and to a lesser degree, the deans themselves. Those specific areas common to the most successful deans Acheson had evaluated were "...genuine interest in each student, sympathy, social ability, a well-adjusted personality, keen intelligence, good heath, and an attractive personal appearance."[27] On the negative side, Acheson found that those deans who were ranked lower or were "less successful" tended to demonstrate characteristics such as "...lack of sympathy, firmness and inflexibility, irritability and unfriendliness."[28] In addition, the unsuccessful deans tended to exert too much authority, mistrusted students, and used punitive rather than constructive methods in disciplinary matters.

By the 1930s, the number of deans of women had expanded significantly since Alice Freeman Palmer and Marion Talbot began their careers at the University of Chicago in 1892. In addition to numbers, the deans of women had established professional standards that they hoped would command respect and attention from the higher education community. As Acheson and others attempted to validate through research studies, the deans of women were competing on an equal level with men and performing well.

But for the most part, the deans of women were still "strangers in a strange land" in that they could not compete on a level field with male faculty, administrators, or even students. The decade of the 1930s marked significant changes in American culture and society and consequently in the work of the deans of women on the college and university campus. Most notably, the economic climate of the country changed radically as a result of the stock market collapse and the subsequent Depression. As economic conditions worsened, so did the access of women to employment outside the home.

By comparison to other fields in which women were employed, the deans of women had been relatively prosperous through the 1920s.

Despite the national trauma of World War I, deans of women had reached beyond the boundaries of the campus to establish a national organization. The National Association of Deans of Women and Advisers of Girls (NADW) was able to command attention not only from those within the profession but from college and university presidents, faculty leaders in various academic disciplines, the leadership of the National Education Association, and even from the deans of men, who were still building their own organization at the beginning of the 1930s.

The NADW had effected an affiliation with the Department of Superintendence of the National Education Association as well as the Association of Collegiate Alumnae (later the American Association of University Women). These connections were certainly a part of the rise to prominence of the NADW. In addition, the fact that the number of women enrolled on college campuses across the United States had reached 47 percent of the total enrollment in the 1920s was also a significant factor in the demonstrating a continuing need for deans of women.[29]

The increasing emphasis on professional organization was by no means unique to the deans of women. In fact, the proliferation of professional organizations across a broad cross-representation of women's occupations was in full force by the 1930s.[30] The continuing application of scientific management on the college campus prompted an increase in specialization and the appearance of new fields and offices. This proliferation of specialization and specialists created new tensions and forced a critical reassessment of the roles of the dean of women and the dean of men. Of greatest significance was the renewed emphasis on the "personnel" movement, which gained increased prominence on the campus, spurred on to a large degree as a direct result of the Depression-ravaged American economy and a desperate need for vocational counseling for students.

Scholars such as William Chafe, Frank Strickler, Lois Scharf , and Nancy Cott have documented the gradual constricting of women's opportunities in the 1930s.[31]These developments marked a major decline in the progress of women from the occupational, social, and cultural accomplishments women had achieved in the earlier decades of the twentieth century. The changes in the 1930s were so dramatic that Berkin and Norton titled their anthology of women's history in the 1920s and the 1930s, *Decades of Discontent.*[32]

Within higher education, women's percentage of enrollment as undergraduate students in colleges and universities had peaked in the 1920s. The number of women obtaining graduate degrees began a long and slow decline in the mid-1930s. Neither of these trends would be reversed until the late 1960s.[33] The deans of women and their national association, the NADW, faced significant challenges in the 1930s, including large numbers of women who could not afford to maintain their membership in the Association, to say nothing of those deans who were dismissed from their employing institutions because of financial exigency.[34]

The Depression gave additional incentive to a variety of frustrations, anxieties and pent-up hostilities. The backlash of public sentiment against working women during the 1930s was more likely propelled by a male labor force frustrated by a wounded economy. Scharf suggested that this anti-woman reaction was further compounded by the resentment of many who felt women had violated the domestic boundaries of female employment.[35]As Patricia Graham and Nancy Cott have noted, the college campus was not immune from such reactions. As the number of women faculty peaked and then began a long and unhalted decline in the 1930s, the deans of women found themselves caught up in the same decline, especially as many colleges and universities faced financial exigency as a result of the Depression.

Despite the exclusion of qualified women from professional opportunities in higher education and the equally pernicious ravages of the Depression on the American campus, the deans of women who remained in place during the 1930s maintained a commitment to their vocation and to the students on their campuses. Surprisingly, in the face of myriad adversities, the 1930s would see the publication of a number of significant research projects within the field of deans of women, including projects on working women, female students, and the profession of dean of women itself. All of these efforts encouraged and augmented the ongoing professionalization of the field.

Nancy Cott, in describing the "grounding of modern feminism," has noted the interest, almost compulsion, of women in the early decades of the twentieth century to embrace professional ideology and practice. In particular, Cott describes the interest in women's professions as

> Neutral scientific standards, based on knowledge rationally and objectively apprehended, constituted an alternative to subjectively determined sex constraints and an avenue to democratization of the power of knowledge. Because of the close relation of the professions to education and service (where women's contributions were acknowledged, to an extent); and because the professions promised neutral standards of judgment of both sexes, collegial autonomy, and horizons for growth, they became a magnet among the potential areas of paid employment

> for women. The structure and ideology of the professions were also forms of occupational regulation. The training, credentialing, licensing, and employment standards established within professional groups (not without internal struggle) from the nineteenth through the early twentieth century were intended to establish and maintain their privileged positions.[36]

In particular, professionals were perceived to be "experts who gave the benefits of their education, authority, and service to society in return for pay, recognition, and influence."[37] Although most professionals were male, what the NADW represented to many of its members was a profession equivalent to many of the male-oriented and dominated fields already in existence on the campus and in society at large. But few, if any, men ever aspired to be a dean of women. So the profession of dean of women was and would be the sole province of women. This exclusivity gave the position of dean a permanence and solidarity for women that few other positions on the college or university campus could equal. As such, the deans of women could exert primary internal control over their field and its representative professional organization. Although the division of labor by gender typically discriminated against women, in the case of the deans of women, separation by gender worked to their advantage, at least in terms of professional opportunities as deans. Because of the absence of men, gender-based constraints within the profession of dean of women were nonexistent.

On the other hand, the complete feminization of the profession created occupational isolation for deans of women; a circumstance that became especially challenging when a dean of women might attempt to expand her goals within the male-dominated collegiate community. Many women who were drawn to the profession of dean had already achieved a significant level of education. A large number of the deans had been faculty members first and many continued to teach even after their elevation to the deanship. So they were accomplished and often competitive people eager to move forward in their careers.

As deans of women, however, these women were often thwarted if they aspired to positions of greater responsibility within higher education. Despite the steady professionalization of their field and their academic orientation and background, deans of women were treated like any other women of the time, professional or not, in that they were "presumed to be emotional, subjective, irrational, and personal" according to Cott.[38] As such, they were not considered capable of leadership and were rarely if ever given encouragement if and when they dared to seek a higher or more visible position such as academic dean or the presidency of a college or university. Although some women did achieve higher office, these instances were almost always to be found in segregated, women's institutions where the direct competition with men was minimal. Thus, the deanship directly connected to "women's concerns" was often the top rung of the career ladder for women in coeducational institutions.

Despite the ongoing efforts on the part of the NADW to professionalize the role of the dean of women on the campus and in connection with other groups, the environment in which such growth was pursued remained a debilitating factor. Although women were allowed more opportunity in various capacities, especially as the number of occupations open to women had expanded geometrically in the 1920s, women's aspirations were confined to a narrow and restricted arena. Women could rise to certain levels within specific fields or professions, such as teaching, social work, nursing, or other traditionally female vocations, but these were quite limited.

Nonetheless, the deans of women could celebrate their accomplishments and the professionalism within their field. In fact, the advances of the deans of women were well-known and acknowledged. In the "Survey of Land-Grant Colleges and Universities," published in the *Report of the U.S. Commissioner of Education* in 1931, the opening paragraph in the section, "The Office of the Dean of Men" stated,

> ...in marked contrast with the clear-cut enumeration of the duties of the dean of women and definition of her functions, the deans of men are apparently groping to discover just what their justification for existence may be. The most illuminating material is found in the proceedings of the National Association of Deans of Men. Here is evident a very masculine sentimentalizing of the work and of the relations with students, which vanished from the discussions held by deans of women a score of years ago.[39]

Such an affront to the deans of men could not go unnoticed, and it no doubt precipitated some of the subdued antagonism toward deans of women and the National Association of Deans of Women that percolated below the surface of the dean of men's meetings. These attitudes may have reflected the stark differences between the deans of women (who considered training and graduate-level preparation to be essential) and the deans of men. The antagonism may have also been misplaced disdain for women in college in general or just professional rivalry. Unfortunately, these attitudes did not encourage any serious efforts at coordinating activities between the NADM and NADW until the late 1930s. Deans of women were eager to cooperate with other professional organizations and did align themselves with the Council on Personnel and Guidance when it was formed in 1934, a bold statement for an association that, at the time, was struggling to keep itself solvent as a result of the Depression and the loss of jobs for women, including deans of women[40]

By 1937, at the close of their annual session, the deans of men considered several final resolutions presented by the Resolutions Committee, chaired by Dean Goodnight of Wisconsin. One resolution introduced read as follows:

> BE IT RESOLVED that in conformity with Dean Park's report and in recognition of the need for greater cooperation with the Deans of Women, this association seeks to arrange at least one joint meeting with the Association of Deans of Women....[41]

Though the motion was discussed on its merits, opposition soon mounted on the news that the deans of women had already voted the idea down in their association. Not surprisingly, the resolution and the possibility of a joint venture were tabled indefinitely.

But cooler heads prevailed and the deans of men and women did arrange to have their representatives, along with members of other professional associations, including admissions officers, vocational guidance groups, and the ACPA, meet in Akron, Ohio, in 1938 with D. H. Gardner hosting. As the NADAM had initiated the meeting, there were eight deans of men present, more than enough to hold off the two deans of women and the representatives from other student service organizations. Nonetheless, the deans agree to keep in touch and make a serious effort at communicating further.

In the same year, 1937, W. H. Cowley, professor of psychology and director of the Bureau of Educational Research at Ohio State University, gave his infamous presentation on the "disappearing deans of men" to the deans of men at their 1937 conference held at the University of Texas.[42] As may be recalled, when asked if deans of women would also be "subordinate" to the personnel coordinator and further, would it be possible for a woman to be promoted to the "top spot?" Cowley's answer was simple: "It seems to me there isn't any reason deans of women shouldn't go up [to the position of coordinator] if they are equal to it. I think we can say that the deans of women are in exactly the same position as the deans of men."

Cowley was hedging his bets. He had chaired the ACE committee that had crafted the Student Personnel Point of View, a monograph that would soon be sent to colleges and universities across the country. He served on that committee with several women who were intimately involved with both the student personnel movement, as well as the deans of women including Thrysa Amos and Esther Lloyd-Jones. Cowley was also aware that the deans of women through the NADW and in academic circles, most notably Sturtevant and Strang at Teacher College, had been strong proponents of the student personnel movement. So Cowley was an unlikely candidate to argue against women holding positions of supervision. In fact, he knew many women who could do just that if they were given the chance. So his response to the question raised by the deans of men was perhaps honest but nonetheless unlikely.

Francis F. Bradshaw, dean of men at the University of North Carolina, had been instrumental in forming partnerships between the old appointment secretaries organization that became the American College Personnel Association, and representatives from the National Association of Deans of Women and other groups, including the admissions officers. Bradshaw was much more conciliatory toward the deans of women, and sided with them in their commitment to the student personnel movement. In fact, his time at Columbia put him in close contact with the faculty at Teachers College, Sturtevant, Strang, and Lloyd-Jones, staunch proponents of the personnel point of view.

There were other deans of men who were favorable toward working with the NADW, and certainly many individual deans of men and women worked with each other on their respective campuses. However, the endorsement by Cowley that either the dean of men or the dean of women might assume responsibility for the coordination of personnel services on campus helped fuel a new sense of competition between the deans of men and women.

In the interim, the push for greater coordination of personnel services on the college and university campuses continued. The meetings of the Council of Personnel and Guidance Associations (CPGA) continued in earnest and eventually gave rise to the American Personnel and Guidance Association (APGA) in 1952.[43] The APGA would serve for some time as the so-called umbrella association for many of the personnel groups, coordinating and at times dictating the directions of several associations and taking on the considerable administrative tasks associated with that coordination including memberships, national meetings, and publications.

While the deans of women, individually and collectively as the NADM, encouraged these mutual concerns, the immediate future of the deans of women was unclear. While women were still attending college in significant numbers, there was still a need for deans of women. However, the dire economic conditions of the 1930s had limited the number of women able to afford college and in turn, began to press the question of the need for deans of women and of men. On some campuses, the activities of the separate deans offices was consolidated into either personnel activities or combined under a dean for students. Examples of both of these options had been cited by Cowley in his 1937 speech to the NADAM meeting, and included Northwestern, the University of California, Berkeley, and the University of Texas.[44] This trend was inevitable, Cowley had declared, adding, "whatever the past has been, I am convinced that a well rounded personnel program . . . will be an absolute 'must' for all institutions of higher education."[45]

THE DEAN OF WOMEN, 1940

American colleges and universities would see their enrollments swell in unprecedented proportions by the end of the 1940s due largely to a staggering increase in the number of students, a population comprised predominantly of returning male veterans. The "vets" accounted for "forty-nine percent of the total college enrollment and sixty-nine percent of all college males enrolled as of 1947."[46] According to Helen Horowitz

> . . . as veterans arrived on the campus following World War II, women students found their interests ignored. Some veterans brought wives with them to college, and this heightened coeds' interest in marriage. Ideological pressures mounted on women to return to the home after college. In this atmosphere, it took a certain independence of mind for a college woman to envision a future career.[47]

It also took a "certain independence of mind" and a strong sense of purpose to be a dean of women. A peak in women faculty had been reached in the late 1930s and very early 1940s and then began a long, slow decline. In the late 1940s, a variety of dynamic changes combined on the college campus to create a "perfect storm" for the deans of women—the number of female students had been in decline since the Depression, the number of female faculty was now declining as well and at the same time, the "student personnel" movement advocated an increase in the services available to students, most of whom were male veterans of World War II. The implementation of the student personnel theory endorsed coordinating all student services under an executive staff member, often a dean. In the mid to late 1940s, in the aftermath of an unprecedented military mobilization around the world, an "executive" or a leader of any type was not likely to be a woman.[48]

These shifts in the composition of the student body, the faculty, and the direction of newly applied theories of student personnel work did not bode well for the deans of women. As college enrollments increased dramatically after World War II, college campuses were desperate to achieve any economy of scale possible. Many campuses doubled or even tripled in size as a result of the G.I. Bill, often in a matter of months.[49] Throughout the 1950s, college campuses, especially the public universities and colleges, struggled to build new buildings, hire new faculty, and create an infrastructure that could manage so many new students. The "old time" dean of men, like Thomas Arkle Clark, would have been overwhelmed by the institutional changes. Even with his platoon of Assistant Deans and clerks, no one could keep an "open door" in the same manner as when many colleges and even some universities numbered their students in the hundreds, not the thousands. The dean of men's message was valid but for another time and another place.

By comparison, early in their evolution as a profession, many deans of women had embraced the cognitive, scientific perspective represented by the "student personnel" movement, whose spirit was captured in the 1937 monograph, "The Student Personnel Point of View."[50] Some of the most influential members of the National Association of Deans of Women, especially the faculty at Teachers College at Columbia, had been staunch advocates of the "personnel movement" since the early 1920s.[51] Deans of women saw the value of data collection, deliberate and conscientious research, and dissemination of research reports. Women had argued for and enrolled in graduate programs as early as 1916.

Prior to the war, both the deans of men and deans of women had been approaching, relatively speaking, common ground. At the NADM meeting in 1937, J.F. Findlay, dean of men at the University of Oklahoma, gave a presentation entitled, "The Origin and History of the Work of the Dean of Men."[51] Findlay opened his 1937 presentation to his fellow deans with a brief summary of the history of higher education in the United States, digressing slightly to describe, with some inaccuracy, the phenomenon of coeducation at the turn of the twentieth century. His research study had been a simple survey sent to ninety institutions. From the results of his survey, Findlay determined that the causes for inaugurating the dean of men's offices were reducible to fourteen top responses. In over one-fourth of the campuses he studied, the prior existence of a dean of women had led to the creation of the office of the dean of men.

Early in 1940, a book *Trends in Student Personnel Work*, by Sarah Sturtevant, Ruth Strang. and Margaret McKim was published by Teachers College, Columbia University, as Volume 787 in the Contributions to Education series. Subtitled "as represented in the positions of dean of women and dean of girls in colleges and universities, normal schools, teachers colleges, and high schools," it was a typical, thorough. and highly detailed Sturtevant and Strang effort.[52]

Trends was the culmination of a massive, four-year undertaking. First initiated in 1936, the research had begun when 1,028 questionnaires were sent to 548 colleges and universities, 227 normal schools and teachers colleges, and 253 high schools across the United States. The collection and analysis of the data and then the writing of the final report took close to another three years to complete.

In their introductory remarks, Sturtevant, Strang and McKim, with a subtle snub to their NADAM research counterparts, noted that,

> ...in order to understand personnel work today in high schools, normal schools, colleges, and universities it is necessary to study both the personnel workers and the institution. An investigation merely of functions is inadequate. Nor is a survey at any one point of time sufficient to give insight into such a mobile vocation as student personnel work. In order to understand its development, trends in the field in the same types of institutions must be studied.[53]

Using previous studies such as Jones' work on deans in colleges and universities, and their own previous work on deans in normal schools and teachers colleges in 1928 and high schools in 1929 as a means of grounding their most recent study in previously established data, the authors discussed the prevalence of deans of women, salaries, academic preparation, teaching, organization and staff, and student personnel functions.

Sturtevant, Strang, and McKim were careful to delineate their findings in several ways. In addition to a standard breakdown by type of institution, colleges and universities, normal schools and teachers colleges and high schools, results from higher education institutions were reported in three groups. Group I were those institutions accredited by the Association of American Universities (AAU). Group II were those institutions accredited only by a regional association Group III were non-accredited institutions. Of particular note was the response rate the questionnaire generated among all institutions. Seventy-three percent of the colleges and universities returned the survey, as did 74 percent of the normal schools and teachers colleges and 79 percent of the high schools, a total of 768 institutions or almost 80 percent of the total 1,028 solicited for the study.

In the early chapters of their book, Sturtevant, Strang, and McKim reported that as of 1936, 89 percent of all colleges and universities responding, or 358 of 400, had a dean of women. Only 42 did not. When broken into groups, 97 percent of the nationally accredited colleges and universities, 89 percent of the regionally accredited colleges and universities, and 75 percent of the non-accredited colleges and universities had a dean of women. Among the normal schools, only 50 percent had a dean of women whereas teachers colleges reported a figure close to that of the colleges and universities, 82 percent.[54]

All of which led Sturtevant, Strang, and McKim (1940) to remark in their conclusion,

> In colleges, universities, and teachers colleges three significant facts emerged about the position of dean of women. It is a well established position; it is a long established position; it is a more highly developed position in the larger institutions and those of higher rank.[55]

Implicit in the authors' analysis was the presumption that those institutions within Group I were the most desirable both in terms of academic integrity and prominence, with Group II relegated to a lesser status and the nonaccredited Group III institutions at the bottom. To enroll women as students and yet not have a dean of women to represent them was considered a significant educational gaffe; one which, in fact, could be traced back to the early ACA recognition of only those institutions which, among other qualifications, met that of having employed a dean of women.

ACADEMIC PREPARATION

In terms of academic preparation, Sturtevant, Strang, and McKim found differences among deans of women in the three Groups. Across all groups, 82 percent held a master's degree. Ninety-seven percent of all deans of women in colleges and universities held at least the bachelor's degree. In the nationally accredited institution (Group I), 98 percent held a bachelor's, 84 percent had a master's, and 28 percent had earned the doctorate. Regionally accredited colleges and universities (Group II) dropped only slightly to 83 percent of deans with a master's degree and 14 percent with a doctorate. At the nonaccredited colleges and universities (Group III), 68 percent held a master's but only 2 to 4 percent had doctorates.[56]

Normal schools and teachers colleges were comparable at the bachelor and master's degree levels but doctorates among the deans of women in either type of institution were scarce. The authors attributed much of this disparity to the differential in salary for the dean of women in normal schools and teachers colleges as compared to the colleges and universities. In particular, the larger, public institutions appeared to have provided the best compensation among the higher education institutions.[57]

SALARY

Given the post-Depression context of their study, it is not surprising that Sturtevant, Strang, and McKim devoted an entire chapter of their book to the topic of salary. Fewer responses were gathered from all of the reporting deans and institutions in this category than any other, perhaps reflecting a general reticence on the part of professionals in any occupation to disclose their salary. Nonetheless, the survey results did reveal several interesting trends.

For one, the authors determined that salaries among deans of women had not recovered to pre-Depression levels by 1937. In a close examination of the issue, 53 institutions or 24 percent reported "a rise [in salary], a drop during the depression, partial restoration." By comparison, only 19 institutions or 8 percent reported a positive trend categorized as "A rise [in salary], a drop during the Depression, restoration to or beyond the highest point reached during the decade."[58]

When Sturtevant, Strang, and McKim compared the salaries of deans of women to figures published by the Office of Education in 1937, they determined that the median salary of the majority of women in the institutions of higher learning fell between that of associate professor and professor. It was higher than that of assistant professor and lower than that of academic deans. Trends in deans' salaries from 1926 to 1936 showed ups and downs instead of a steady increase or decrease. On all educational levels, the deans' salaries showed a common pattern of decrease following the Depression and "an increase by 1936 toward but not quite reaching, the pre-Depression level."[59]

Another conclusion reached regarding salary was that there were great disparities between deans of women, even within similar institutions. As Sturtevant, Strang, and McKim noted, "In this study, one dean was serving at a salary of about $700, whereas another on the same educational level received about $7000."[60] Such discrepancies were even more obvious across institutional classifications, and not always in the direction one would have assumed; for example, the authors noted "some able high school deans would be forced to accept a lower salary if they changed from a high school to a college position."[61]

Overall, the average salary for all levels of dean was $2700. Based on a median of $2779 (roughly $46,000 in 2010 dollars) for deans in colleges and universities in 1936, deans of women had lost ground from the median of $3000 reported by Jane Jones almost ten years earlier. A comparable drop but with a lower median salary was noted in normal schools and teachers colleges. In their concluding paragraph to the chapter on salary, Sturtevant, Strang, and McKim echoed a common lament: "Everyone concerned with the welfare of students should put service first and salary second, but justice demands that the one be fairly commensurate with the other."[62] As a small group of professional women, the deans of women were comparable to a 1933 American Woman's Association survey of 1350 "white-collar, relatively highly educated women," as noted by Solomon with "a median income of $2,400 . . . at a time when a large number of American families earned less than $1,000."[63]

STRUCTURE AND ORGANIZATION

In a chapter on "Organization and Staff," Sturtevant, Strang, and McKim assessed the trends toward a change in the administrative schema of the institutions polled. In particular they were looking for an indication that the position, dean of women, was being altered by changes in organizational structure. Through a "study of the entire questionnaire," the authors noted that 66 percent of the deans who responded had in fact reported a change on their campuses "toward a centralized program" of student affairs or student services. In most cases, centralization meant that the dean of women would report to a dean of students or personnel director, rather than directly to the president of the college or university. A much smaller number, just over 12 percent, reported a change toward "decentralization" or a dismantling of the institutional bureaucracy. In recognition of the vagaries of institutional organizing, the authors also reported that 16 percent had reported changes that "could not be classified in either direction" and another 4 percent reported "what appeared to be regression or contraction in personnel functions"[64]

Sturtevant, Strang, and McKim noted several trends, which they acknowledged as "impressions more than generalizations." One of the trends was the slow but clearly discernible increase in faculty who were involved in the "student personnel" area, most often reflected in faculty committees set up to supervise certain functions as well as a reported increase in faculty advisers. The research results documented a significant percentile change across the ten-year period. In 1926, only 23 institutions or 10 percent had referred to faculty advisers whereas in 1936, the number had increased to 115 institutions or 34 percent.[65]

Sturtevant, Strang, and McKim also identified four types of reorganization that characterized student personnel from 1926 to 1936 and beyond. One was changes made in the person responsible for coordinating the program. The gender of this person was not clear because of title, such as dean of students or director of personnel. A second change was that a committee would replace a single individual who had been in charge. In one case cited by the authors, the dean of women was replaced by a committee on student affairs for women, as had been done when Marion Talbot retired from the University of Chicago. In another case, a personnel council was appointed consisting of the president and vice-president of the university, the deans of women and men, the examiner, director of the health service, junior deans, two secretaries, and three professors. The third configuration was an increase in staff without significant changes in organization, such as the addition of assistant deans or a director of personnel. The fourth change noted in personnel functions was expansion in the dean of women's office and increased centralization so that for example, on one campus the dean of women was now the "moderator of all women's activities."[66]

In a review of those working in the personnel area, Stutevant, Strang, and McKim found that the academic dean was named most often in response to a question asking what institutional officers performed personnel functions on campus. In fact, the academic dean was second in frequency only to the dean of women. The third officer most frequently named was the dean of men. These responses reflected the nature of the personnel office in the 1930s, when record keeping for each student, along with academic advisement and vocational guidance, were central issues.

These findings gave Sturtevant, Strang, and McKim the opportunity to editorialize that the "dean of men should work closely with the dean of women so that the personnel work with men and women students could be properly integrated. In some institutions, the offices of dean of men and dean of women perform parallel functions with very little coordination of services. Such an organization is unfortunate in that it prevents a natural fusion of the activities of men and women students."[67]

STUDENT PERSONNEL, NOW AND FOREVERMORE

In a chapter on student personnel functions, the authors reported that those personnel functions most frequently mentioned in 1936 by deans of women were counseling students regarding educational, financial, and personal problems; work with student groups; vocational guidance; placement; housing; and work on committees or of an executive or administrative nature in conjunction with the dean of women's office. Of these, those performed most often outside the dean of women's office were admission of students, vocational guidance, and placement after graduation. In essence, the report demonstrated that although an increase in services to students may have occurred between 1926 and 1936, with many of them re-named as "personnel services," in many cases, the person responsible for the performance of those functions was still the dean of women and/or her office.[68]

Perhaps anticipating future changes in the area and clearly encouraging a redefinition of the work of the deans of women, Sturtevant, Strang, and McKim in their conclusion called for a better understanding of the contributions of the dean of women and the dean of girls "[which] requires an evaluation of the efficacy of their work in terms of desirable changes made in students. There is also a need for clarification of the meaning of personnel work and a common recognition of certain fundamental goals and purposes."[69]

Later in the same section, the authors acknowledged that the deans of women would need new skills in certain areas in order to continue to be successful. "It may be that deans of women and deans of girls have been too concerned with details of housing and social activities.... Deans of women or deans of girls functioning as the sole personnel specialist in an educational institution must acquire the psychological background and technic [sic] of work within individuals that will win them the respect of the faculty and students."[70]

However, the authors were not above finding fault with the new emphasis on "personnel functions." Specifically, they noted that

> ...other personnel officers, while admirably trained in psychology and psychometrics, have neglected the important area of group activities.

> In their concern for the scientific study of the student, they have not made provision for the developmental and therapeutic values of group discussions, dramatics, sports, games, committee work, and social events.[71]

One of the earliest monographs published by the American Council on Education (ACE) on behalf of its Committee on Student Personnel Work and following the 1937 Student Personnel Point of View (SPPV) was a monograph on "social preparation." Written by Esther Lloyd-Jones, one of sixteen contributors to the Student Personnel Point of View and a faculty colleague of Sturtevant, Strang, and McKim at Teachers College, *Social Preparation* first appeared in September 1940, several months after Sturtevant, Strang, and McKim's *Trends in Student Personnel*. In her preface Lloyd-Jones explained that "social competence has to do with the responsibility that student personnel programs in colleges and universities bear for the development of socially competent college graduates."[72] In addition, she noted that she had intentionally

> ...used the titles 'personnel worker, 'director of personnel', and 'dean of students' ('dean of men', 'dean of women') synonymously, although those who were first called directors of personnel have until recently tended to concern themselves with technical functions such as testing, record-keeping, and placement, and little with the social program. The best deans of students, on the other hand, have traditionally been responsible for much of what we here describe as the social program.[73]

In the same section, Lloyd-Jones called attention to the issue of reconciling the titles of the various institutional officers concerned with student life. Specifically, she asserted that,

> ...although distinctions between deans and personnel workers are sometimes still implied in the personnel field, the former distinctions are no longer justified, a great deal of the cause for confusion is now rapidly disappearing, and it seems fair to use the titles dean of students (dean of men, dean of women), and personnel officer synonymously in this brochure.[74]

In her preface, Lloyd-Jones also thanked her advisory committee, a group that included F. F. Bradshaw of the University of North Carolina and Sarah Sturtevant of Teachers College, for their assistance in preparing her brochure. In many respects, the personnel movement and the deans of women were reaching common ground in 1940. As Bradshaw had noted at the NADM conference in 1939, the deans of men were coming closer to an acceptance of the "personnel point of view," albeit at about the same speed as a mongoose approaches a cobra.

In one sense the deans of women, led by female faculty members at Teachers College as well as leaders in the NADW, such as Irma Voight, Dorothy Stimson, Harriet Allyn, Thrysa Amos and others, embarked on a trail familiar to them, a trail marked by the inevitability of change. The deans of women were, for the most part, highly educated and dedicated women. Their predecessors had been steeped in the feminist tradition of the nineteenth century "woman movement" and survived the "flapper" era of the 1920s. Deans of women had learned to accommodate the austerity of the Depression and struggled to emerge with their positions and their national organization intact. Now they faced another change, a change that threatened the structure of their profession through the personnel movement.

As a profession, the deans of women were well prepared to contend with such change. The deans had well-defined national organization as well as state and local organizations. The publications of the NADW including the new *Journal of the NADM*, published quarterly, had been a success. The deans' long-standing commitment to academic preparation for their professional work equipped them for continuing education and study. This academic preparation also encouraged and supported innovation and ongoing research, such as Sturtevant, Strang, and McKim's *Trends in Student Personnel*, as well as Lloyd-Jones' new monograph. Consequently, the expanding focus on the student personnel movement was not perceived as a threat but rather as another challenge, perhaps even an opportunity. Indeed, when Sturtevant, Strang, and McKim urged that deans of women needed to "acquire the psychological background and techniques of work with individuals that will win them the respect of faculty and students," they were speaking to an audience of colleagues, friends, and former students.[75]

The deans of women were very different from other women in the 1930s and 1940s. Most had earned graduate degrees, survived the economic purge of the Depression, and earned a reasonable, even upper-middle class salary[76] As described by Barbara Solomon and Nancy Cott, some professional women, as were the deans of women, could have been seen as isolated from, even insensitive to, other working women, particularly given the nature of their positions and their extensive education.[77]

The deans were members of a professional elite but they existed in isolation even from each other. There was typically only one dean on a campus and few deans of women were on campuses in close proximity to each other. At the national level, even the NADW could only count a thousand or so deans in the entire country.[78] Although not atypical for their age group, many of the deans had remained single, dedicating themselves to career and their students. As such, the deans were further estranged from traditional female roles of wife and mother.

Even though these trends would change slowly over the years, the idea that a woman could combine marriage and full-time employment was antithetical to commonly held assumptions in the United States in the 1920s, 1930s, and in the pre and post-war 1940s. Even as sophisticated a group of collegiate alumnae as Radcliffe graduates educated between 1883 and 1928 were less than unanimous in their outlook on marriage and career. Barbara Solomon notes that in a survey of 2000 alumna of Radcliffe who responded to a 1928 questionnaire, three-fourths thought it possible to be married and work; however, only one-third who answered the question gave "unequivocal, positive answers whereas another 40 percent qualified their positive answers." Solomon also noted that when asked about the possibility of combining career, marriage, and children, the numbers declined further to 49.3 percent who answered "either positively or hopefully."[79]

When the questionnaire was repeated in 1944, the numbers dropped to "only 18.0 percent and of those who were hopeful 39.3 percent" with regard to combining marriage and career. In attempting to interpret the decline from the 1928 to the 1944 survey, Solomon has hypothesized that three explanations were possible: (1) the Depression; (2) disappointment that suffrage had not given women equality in the professional sphere; and (3) the impact of World War II, which left women eager to return to domestic "normalcy."[80]

In the preface to their book, *Decades of Discontent: The Women's Movement, 1920 to 1940*, Jensen and Scharf have noted that although "single women in the late 1800s and early 1900s had been able to associate with each other personally as well as professionally without reproach, by the 1920s,...spinsterhood and celibacy became suspect."[81] If heterosexual relations were normal and healthy, then women who lacked such experiences became, by inference, deviant.

> ...If single women vented their anger against their working sisters during the Depression, working women could return the diatribes in

> kind. Why dismiss teaching wives and mothers, they asked, and turn schools into havens for bitter, sex-starved old maids who were unfamiliar with "normal" life-styles?...Single women were suspect; married, they were deprived of the same-sex camaraderie that could soften the discrimination, hostility, or neglect they encountered....[82]

It is difficult to think of the deans of women as "bitter, sex-starved old maids" who were not familiar with "normal life-styles" but it is impossible not to speculate as to how much of that perception of the deans of women was muttered in the hallways and classrooms of the college campuses. As a group, deans of women were bright, highly educated, and dedicated to their profession. Though positive, these qualities could also result in a perception of the deans as individuals who were spinsterly, anachronistic, and with whom it was difficult to work, particularly as they championed the concerns of women on male-dominated campuses.

THE DEANS OF WOMEN IN THE POST-WAR YEARS

In her dissertation completed in 1952, Louise Spencer replicated much of the research done in 1936–1937 by Sturtevant, Strang, and McKim. Not surprisingly, Ruth Strang was Spencer's major adviser and even signed the introductory letter that accompanied Spencer's survey instrument, a questionnaire sent to 677 institutions. Spencer collected completed surveys from 472 colleges, universities, normal schools, and teachers colleges, a response rate just over 69 percent. She compared her results against those compiled by Sturtevant, Strang, and McKim as well as the earlier studies by Sturtevant and Strang in 1928 and Jones in 1928. As a result, a longitudinal picture of the deans of women was linked together across several decades.

Spencer identified several trends among the deans of women that had emerged since Sturtevant, Strang, and McKim had finished their report, a period that included the years after World War II. Spencer was able to determine that among those institutions responding to her survey, 56 new positions for deans of women had been created, continuing the increase Sturtevant and colleagues had identified in their data collected in 1936.[83]

Spencer also determined that the trend toward an increase in academic preparation among deans of women had continued, despite some faltering during the war years. In particular, Spencer noted that those deans who had entered the field between 1942 and 1948 tended to have lower academic qualifications than the total group, probably due to the dearth of potential applicants during the war. However, by 1947, fewer than 1 percent of the deans of women did not have a college degree, and 13 percent had earned a bachelor's degree. Even more significant, 68 percent of the deans had earned the master's and 18 percent had doctorates.

By 1947, the median salary was $3633, an apparent increase of almost $1000 since 1936. But Spencer realized this increase did not cover the concurrent increase in the cost of living, which had also risen since 1936. She concluded that in reality the deans as a whole had sustained a loss in real dollars as salaries ranged from $1200 to $8799. Median salaries for deans with a bachelor's degree were $2850 while those with a doctorate reported a median salary of $4350. The number of deans of women who still had classroom teaching responsibilities had dropped significantly from 1926 when four-fifths of the deans taught, as compared to two-thirds in 1947. However, Spencer noted that "the higher the academic qualifications of the dean of women, the more likely she would be teaching in all institutions."[84] In addition, Spencer found that for all deans of women who taught, the average course load had declined by one.

Spencer also found that whereas in 1936, less than 2 percent of the deans of women reported to a "coordinator of personnel services" in previous years, by 1947 that figure had increased to 12 percent. Although the deans were "generally convinced" that personnel services held "significant potential value," especially in the areas of personal counseling, developing personal-social responsibility and orientation, there were problems in "nearly every area of personnel service." These problems were ones of delivery, especially as "influenced by facilities, staff, organization [and] coordination of services." Of particular concern, Spencer noted, was

> ...the need for greater understanding by faculty members of the personnel program...[and]...the heavy concentration of responsibilities in the office of the dean of women which may limit the quality of services for students and restrict her [the dean's] activities for personal and professional growth.[85]

Given the time frame during which Spencer collected her data, a period immediately following the end of World War II, the deans of women, along with other administrators and faculty, were in the midst of a temporary but serious work overload. In particular, one of the primary concerns of the deans of women was the development of more programs on the topic of marriage and family living.

"COORDINATION": A NEW NEMESIS

In completing the interpretation of her survey research, Spencer was privy to the beginnings of another new and significant trend in the history of deans of women; a trend which would prove to be irreversible. In a subsection of her dissertation titled "The future of the trend toward coordination," Spencer commented,

> One of the trends noted in the 1936 survey was in the direction of a larger number of directors of personnel. No single figure in the current survey indicated the extension of this trend toward coordination in 1947–48.... It is obvious that within the last eleven years the trend has been accelerated. Moreover, it seems probable that the next decade will mark further extension of this trend. A number of deans of women indicated that changes were already in progress, or that plans were afoot for greater coordination within the next year or two, or that the need for unity in the program was evident.[86]

Spencer noted that many of those deans of women had been subsumed under a coordinator or director of personnel services but the deans appeared to be satisfied with the result as they were able to continue their previous work unabated. On the other hand, those deans who had been reorganized under an "administrative dean" (in which case a differentiated program for women students was not included because "it recognized neither psychological nor sociological differences between students of different sex") (p. 123) were very critical of the changes.

Spencer examined the issue of reorganization extensively and included some quotes from many of the affected deans of women. Some deans felt that the reorganization of the dean's offices would result in better service to all students. "I am now under the Dean of Men or director of Student Affairs, he is now called.—I like this new system better. The work is coordinated and I can take all my problems to him. Before this time I had to go to the President, who was too busy." Another dean of women responded in the same vein. "The new position which has been added last year is Dean of Student Affairs. He acts as a coordinator of all personnel activities and his coming has been a wonderful help to this office."[87]

Other women were not as pleased with the new arrangements. "I seriously object to being just another secretary in the Personnel Department; working out research material, submitting it for use, and finding it presented to the Administration as the work of the Dean of Students alone.... Another disgruntled dean wrote on her questionnaire, "We are in the process of extinction. The office [dean of women] will doubtless be discontinued upon my retirement—. No change [in title] but title is very empty honor. Almost no functions left in my office since—."[88]

Spencer wrote passionately about some of the changes she had encountered in reviewing the responses to the restructuring of dean's offices. The sharpest reactions concerned the removal of deans of women from policy-making boards of their institutions. Although there were exceptions, in general the "Administrative Dean" or the "Dean of Students" or the "Director of Personnel Services" represented all personnel interests and needs of all students to policy-making boards, including the president and community groups. No longer considered one of the top administrators, deans of women did not represent the school with the same authority they had become used to in the preceding years.

The crux of the issue, as expressed by the deans of women was not a matter of personal prestige, although that may have been the case for some, but the concern that the needs, problems, and interests of women students could not be adequately understood or represented by a man. Though often well-intentioned, a dean of students who was a man was typically far removed from the problems experienced by women. Even women faculty would lose some of their voice if the dean of women was no longer in office.

The trend on college campuses that concerned Louise Spencer was also rippling through other occupations in post World War II America. Just as Rosie the Riveter was squeezed (some would claim pushed) off the assembly line, so was the dean of women nudged out of the decision-making structure on campus. Formerly a significant symbol of the female presence on the campus, the office of the dean of women was "restructured" out of the administrative organization of many college and universities in the late 1940s and early 1950s.

As Coomes, Whitt and Kuh have noted, "As men returned to reclaim their jobs, staffing in student affairs divisions mirrored the return to a male-dominated academy. At many institutions, the roles of dean of men and dean of women were consolidated into one: "the dean of students."[89] To the dominant gender on the campus, the dean of women was seen as an anachronism, out of time and out of place. Consequently, on many campuses, the office along with the officeholder was slowly pushed into obscurity as the college population swelled with male student veterans.

In a biographical sketch of Kate Hevner Mueller, dean of women at Indiana University from 1937 to 1947, Coomes, Whitt, and Kuh pieced together her reaction to "reorganization." Mueller, having been relieved of her post as dean of women during a "restructuring" in which the dean of men was made dean of students, was reassigned to the campus counseling center. After two years in that position, the president of the university and Dean of Faculties offered her an assistant professor position. Still angry over her dismissal as Dean of Women and the inherent disrespect for her abilities, Mueller responded, " 'I was an assistant professor before I came here and I have no intention of being one again'; and she got up and walked out on them."[90] Mueller was offered an associate professorship a few days later, a position she did accept, and spent the next twenty years teaching, writing, and conducting research. Not many deans were capable of the response of a Kate Mueller, who has been compared to female leaders in student personnel such as Strang and Lloyd-Jones.

Despite the tremendous increases in women been employed outside the home during World War II, when the war ended, women were pushed out of the labor force. Scholars such as Helen Horowitz, Barbara Solomon, and William Chafe have concluded that enormous social pressures were brought to bear on young women to return to the domestic roles of wife and mother. Pamphlets produced by the Secretary of War in 1943 declaring, "You're Going to Employ Women" to encourage business and industry to hire women during the war were replaced by 1947 with a determined effort to exclude women just as swiftly. To many, a woman's place was still in the home, where she posed no threat either to men or to other women.

> Chafe examined the rates of female employment after World War II and found that, women fell from 25 percent of all auto workers in 1944 to 7.5 percent in April 1946....Overall, females comprised 60 percent of all workers released from employment in the early months after the war and were laid off at a rate 75 percent higher than men.[91]

SUMMARY

As the decade of the 1940s moved forward, the influence of World War II on society in general and college campuses in specific was enormous. Much of the pattern of American higher education changed forever as a result of massive infusions of federal money to students through the G.I. Bill and to institutions for buildings and research. The populations of the post-war campuses would swell almost to the bursting point, an unprecedented phenomenon with most of the increase directly due to male veterans who saw a college education as the answer to their dreams of prosperity and happiness.

Some of the changes would prove of great benefit to American society. In particular the college educated veterans allowed the transition of the country from the war effort to a peacetime economy in a relatively short order while concurrently avoiding the onslaught of a second Great Depression. During the same time period, the "baby boom" was inaugurated as young men and women, no longer confronted with the uncertainty of war, dated, mated, and gave birth to an unprecedented population boom.

What is still being calculated, however, is the toll extracted by the post-war discrimination and oppression of American women. Although women had contributed to the national defense and economy with long hours of labor and toil through both voluntary and paid employment, they were not among those who directly benefited from the post-war prosperity. A variety of messages were transfused into the body of American culture through "scientific" studies and the popular press. Chafe cites two particular examples, a book by Philip Wylie published in 1942 titled, *Generation of Vipers*, and *Modern Woman: The Lost Sex* by Ferdinand Lundberg and Marynia Farnham, which appeared in 1947.

Wylie, Chafe notes, specifically attacked what he perceived to be the mother-worship fetish in the United States. Not content to simply denounce the practice, he wrote in a style "designed to provoke and titillate rather than persuade." Using phrases such as "she plays bridge with the voracity of a hammerhead shark..." and "I have researched moms to the beady brains behind their beady eyes and to the stones in the center of their fat hearts..."[92]

Lundberg and Farnham focused on the "scientific" side and claimed that all women with a feminist point of view had suffered from bad experiences in their childhoods, often precipitated by a terrible father whom they constantly attacked in adulthood by castigating other men. In their book, the authors proposed a definitive program designed to rectify the stature of motherhood and the joys of being a wife. Chafe notes that

> ...they urged a government-sponsored propaganda campaign to bolster the family, subsidized psychotherapy for feminist neurotics, cash subsidies to encourage women to bear more children, and annual awards to mothers who excelled at child-rearing.[93]

In the late 1940s and early 1950s, these attacks on women by Wylie and Lundberg and Farnham generated considerable attention and gained increasing credibility. Similar articles and sketches of aggressive, "unwomanly" women appeared in much of the popular print media. In short, the social ruse was that women belonged at home and not in the workplace; to argue otherwise bordered on a perversion of the "natural" order. Even the new medium, television, soon perpetrated the same message, as popular television shows of the 1950s portrayed the ideal family as seen on "The Ozzie and Harriett Show," "Leave It to Beaver," and the ironically named, "Father Knows Best," programs that always showed Mom at home in the kitchen, baking cookies, and beaming over the family dinner in a dress and high heels.

On the college campus in the late 1940s and early 1950s, women soon found themselves in a distinct minority, a status that would remain in place for the next 25 years as many women were encouraged to either avoid college or attend classes only long enough to earn an "M-R-S" degree. The college and university enrollment of women, once as high as 47 percent in the 1920s, declined to 31 percent in 1950. Female faculty fell from 28 percent in 1940 to 24.5 percent in 1950, dropping to 22 percent by 1960.[94]

Although the number of higher education institutions increased, especially among community colleges and adult education programs, in "... this critical period between 1945 and 1956, women as a group were handicapped by the male influx into academia." Solomon has also argued convincingly that

> fearful of the continuing changes in work patterns and in expectations of women trained during and after the war, educators, relying on studies in psychology from a Freudian perspective, again succumbed to curricular arguments for a feminine education. The proponents cited wartime and postwar surveys of college women to justify making liberal education for women different from that for men....Such surveys gave further evidence that women in a wide spectrum of colleges wanted courses on family life....Overreacting educators interpreted these responses to mean that women wanted solely a domesticity for which higher education unfitted them.[95] (pp. 191–192)

The personnel movement or "student personnel" increased services to students in large measure and even began to achieve Cowley's goal, as stated in 1937, of coordinated services. However, such services were increasingly organized under a male administrator and accomplished within the context of a predominantly male campus. In many respects, the organized personnel services, based on "scientific" studies and assessed on "scientific," measurable indices, were ill-conceived to measure the dynamic effects of college life on subgroups of the student population. Personnel services professed equal treatment of all students but "all students" was more likely to refer to the majority population of white males rather than the divergent populations, such as women. These conditions were accentuated by the all-too-common exclusion of the dean of women from the administrative hierarchy[96].

As Solomon[97] has described, campus life for women has been grossly misinterpreted when examined from a gender-biased, male perspective. Student personnel practices in the late 1940s and early 1950s were often based on tests, measures, surveys, and the like, in essence, quantifiable, easily organized measures that appealed to beleaguered administrators who were confronted with huge, unprecedented numbers of students after World War II. As such, student personnel had not only intellectual but esoteric appeal. Here at last was a deliberate, unified means of recordkeeping and classification; it was an outcome-oriented theory and process combination that not only made sense to presidents and boards of trustees but also satisfied the constant query of "what is it you do?", a question that had plagued deans of women and men for decades.[98] Deans of women, through the logic and science of the "personnel point of view," conceded their instincts and intuitions for qualitative as well as quantitative processes and joined the movement with a certain measure of blind faith in the academic scientist the personnel theories extolled.[99]

However, when personnel theory and practice emerged on the post-war campus, it did so under the joint sponsorship of men and the military. This post-war environment was strongly influenced, even created, by the desperate need of many powerful forces and people in America to maintain a position of strength and dominance in the post-war world. In this case, power was a male norm; it meant traditional, accepted roles for men and especially women.[100] (Women belonged in the domestic sphere. Men were best suited for the public roles of workers and leaders. As a result, those engaged in rebuilding America sought to exclude women from positions of leadership and high visibility. Deans of women, often single, childless, highly educated and personally assured, if not occasionally obstreperous, were deviants in the new culture and thus, unnecessary. In the rush to reconstruct America, an America composed of marriage, families, children, men at work, and women at home, deans of women were an obstruction to be avoided, eluded, or eliminated.[101]

Spencer's observations in the late 1940s and early 1950s about the lack of a voice for women on the college campuses were poignantly accurate. Spencer had expressed concern in 1952 about the impending dismantling of the positions of dean of women, and posed the query, how "shall the needs and problems and interests of women students be represented..."[102] without deans of women on the campus? The question was to remain unanswered for some time.

Despite this apparent advantage, it was the deans of men who would emerge victorious. As the G.I. Bill fueled the greatest enrollment surge in the history of American higher education, deans of men, not deans of women, captured the "top spots" Cowley had described. Men, not women, were promoted to be "the deans for student personnel services" or "deans of students" in the late 1940s and early 1950s. A strong push toward science and scientific training in many "research" universities after World War II did not help the deans of women, despite their proclivity for the "science" of the personnel movement.

A strong case was made by a simple equation. The deans of men were promoted to positions of authority and leadership because they were men. Military experiences and the overwhelming male presence on college campuses after WWII restored the "male" college culture to America, after it was briefly lost in the 1920s when women accounted for 47 percent of all undergraduates in higher education.[103] In the male culture that dominated higher education in post-war America, gender trumped personality and professionalism.

Although the number of higher education institutions increased, especially community colleges and adult education programs, in "... this critical period between 1945 and 1956, women as a group were handicapped by the male influx into academia."[104]

Deans of men, like D.H. Gardner, had served in the war, and shared the experience of the largest fraternity, the military. Institutional cultures in the post-war colleges and universities rewarded collegiality and fraternal experience, both formal and informal. The bonhomie of victory in war, and achievement in civilian life carried the day. Massive enrollment of men drove the percentages of women on campus to new lows, further decreasing the need for deans of women and reinforcing the need for male authority and leadership.[105]

As William Chafe has argued, American society wanted desperately to return to a more traditional constellation—often at any cost. Popular media, institutions, and certainly colleges and universities reflected these pressures and preferences.[106] To a large extent, these pressures demanded that men be in positions of leadership while women returned to the domestic sphere, as in the eighteenth and nineteenth centuries. Deans of women, despite their successes in defining a research agenda, professional journals, and the creation of a profession, were interlopers in a male culture, and became strangers in a strange land where men prevailed.

While the National Association for Deans of Men ceased to exist in 1951, the deans of men lived on as deans for students and directors of student personnel. On many campuses, especially smaller, private, liberal arts colleges where tradition maintained the status quo, deans of men and women persisted into the 1960s. But for the most part, the old concept of deans of men and their corollary, the deans of women, ended as World War II stopped and the post-war years of the 1950s began.

8

The Demise of the Deans of Men and the Rise of the Deans of Students

The War Years, the 1940s

Among the changes experienced by the NADAM entering the 1940s was a surge in membership, beginning with a record-setting registration for the twentieth anniversary conference held in Madison, Wisconsin, in 1938, as 164 members and guests attended. Repeating his role from 1919, Scott Goodnight of Wisconsin was the host for the celebration. Attendance was again over 100 in 1941, when the conference was held in Cincinnati, and in 1949, a record turnout of 217 people attended the conference in Highland Park, Illinois.

But the attack by the Japanese military on the U.S. Naval base at Pearl Harbor, Hawai'i on December 7, 1941, would change America, and with it, the national association and the role of dean of men. Despite Cowley's warning about disappearing deans, the huge influx of veterans onto campuses after World War II created a demand for many more student deans. In fact, the increase in the NADAM registrations was due to many more "assistant" deans of men who began to appear after the war, paralleling the increases in student enrollments in the late 1940s.[1]

But the other trend affecting the NADAM in the 1940s was the changes brought about by the war. D. H. Gardner, dean of men at the University of Akron, had served as secretary of the NADAM from 1933 to 1937, and as president from 1938 to 1940. His book, *Student Personnel Services* (1936), was published in the midst of rampant NADAM opposition to the personnel concept.[2]

During the war, Gardner served as a lieutenant colonel in the U.S. Army. Speaking to his friends and colleagues at the 1944 NADAM conference on "The Army Specialized Training Program and Its

Implications for Current and Postwar Personnel Work," Gardner was cautiously optimistic about life after the war. He challenged the deans to assume leadership for post-war educational programs. "To be quite candid, I have become alarmed at the failure of educational leaders to sense the influences which the war is having upon institutions of higher education ... [and] to the future of an American society which will have suffered a hiatus in educated citizens."[3]

Gardner cautioned the assembled deans that the students returning to the campus after the war would be significantly different from when they left. The need for social, vocational, and personal readjustment of all students would be a challenge to all the deans and their institutions. The post-war era could radically restructure higher education. His words, an echo of Cowley's 1937 address, were becoming familiar to the deans of men.

> The number one problem for deans to attack is—the establishment of a student personnel program in all its ramifications.... Many viewed with alarm the rise of personnel experts and technicians. Lengthy discussions have occurred.... Some felt the Dean's work was being "stolen" by scientific "nuts." Others thought there was no need for individualized counseling. Whatever the past has been, I am convinced that a well rounded personnel program ... will be an absolute "must" for all institutions of higher education.[4]

Many of the changes Gardner predicted came to pass. American society longed for "normal" lives in peacetime and many insisted on a conservative array of social, vocational, educational and personal adjustments. The conservative nature of America in the late 1940s and early 1950s along with the huge enrollments spurred by the Serviceman's Readjustment Act of 1944 changed the shape of American higher education and the work of deans of men. Cowley's 1937 address on the "disappearing deans of men." Turner's surveys of 1936 and 1939, and Gardner's 1944 speech were harbingers of changes to come.

The arguments persisted. The "cult of personality," the notion of the dean of men who was born, not made was not yet dead, but it was mortally wounded. The combination of another enrollment surge after the war, a new emphasis on bureaucracy as efficiency, and the weight of the personnel movement were too much to resist. At the NADAM conference held in St. Louis in 1951, after two years of debate, a motion was passed to change the name of the professional association representing the deans of men. The National Association of Deans and Advisers of Men, created in 1919 and officially made the professional association representing the deans of men in 1920, became the National Association of Student Personnel Administrators.[5]

CHANGE IS INEVITABLE

But the change in name did not come easily nor did it address all of the concerns that affected the NADAM in the 1940s. In fact, the NADAM was an organization in the midst of many changes, not just in name, but in terms of mission, composition, and direction. At the 1949 conference held in Highland Park, Illinois, the president of the NADAM, J. H. Newman, Dean of Men at the University of Virginia (known to his intimates as "Foots"), called the deans' attention to comments he had collected after the 1948 NADAM conference. In particular, "Foots" Newman was concerned about some of the less-than-exuberant reactions of some of the new members, many of them younger men who had only recently begun to attend the NADAM meetings. Newman disclosed that he had written to a number of the members after the 1948 meeting in Dallas and asked them to give him their "unvarnished opinions" about the NADAM.

Among the comments Newman received were the general sentiment that the "Association is in the hands of a few to the exclusion of others." Newman then read excerpts from some of the letters who had responded in kind about their concerns. One letter writer complained that the NADAM was "run by a small group of insiders." Another letter writer was more eloquent. "I found a few old main limbs of the tree that objected to some of the new sprigs asserting their right to their particular share of the sap and their particular place in the sun, and I also found that these young twigs were just strong enough to hold their own."[6]

Newman also noted that the new members had not felt especially welcomed at the Dallas meeting. One comment in particular expressed the sentiment that "One has the impression that NADAM prizes venerable experience and prestige. If men are honored who have served in personnel administration in NADAM for many years, why are not the few men made over as entering a distinguished and worthy profession and association?" In addition, there were comments about the program at Dallas. For example, one letter indicated, "the meetings and actual sessions are informal to a fault. This makes discussions so general that their value is limited to those who are actually engaging in practices well beyond the professional level of the discussions themselves." The same letter indicated, "I should like to see the presentation of papers based on experimentation and research as well as upon the authority and status of the person presenting the topic."[7]

These comments spoke directly to the informal nature of the NADAM meetings, a culture the deans of men had encouraged since the very first meeting in 1919. For the most part, many of the deans knew each other and were comfortable with each other. These annual meetings were often full of anecdotes and stories with much less emphasis on research or related activities that might have been expected at a professional meeting. The "club" atmosphere was not a mistake or accidental, it was how the deans of men had shaped their time together for many years. But clearly, it could be off-putting to new, younger men who were emerging from academic settings and the recent experiences of a graduate education.

Newman also cited letters chiding the NADAM organizers for omission of some key topics. One letter stated, "I should like to see the younger group given the opportunity to show the old folks something by way of personnel guidance, counseling, or what-have-you." Another comment was about the relationship between the NADAM and the American College Personnel Association (ACPA). The comment was, "I have only one observation of the Dallas meeting... That deals with the very controversial question of the relationship to ACPA. While I can understand that there are some very real differences in point of view and approach, it seems to be the fact that the two groups are seemingly drifting apart. I regret to see it become necessary for a choice to be made between the two organizations which fundamentally have more in common than they have separating them."[8]

One other letter writer expressed his sentiments, which also framed the issue at hand for the NADAM and the organization's professional role in the future.

> It is obvious that the Association is experiencing some growing pains which reflect the changes in personnel organization going on at many institutions. The majority of membership is still made up of deans of men who are responsible only for men and frequently do not have duties along lines of such services as vocational guidance (except in a general way), health services, etc. which involve all students. On the other hand, the number of deans of students is on the increase and since these tend to represent the larger schools, they tend to be in places of leadership. This makes for some conflict within the organization [referring to the NADAM] (however friendly it may be) and makes a focus of objectives somewhat difficult. This was reflected by the committee's report and the discussion on "Survey of Functions of

Student Administration for Men", etc. and the proposals and discussion on change of name for the Association.[9]

Newman also shared responses from NADAM executive board members to the letters he had selected. Generally, the consensus was that indeed a change was affecting the NADAM and it was important to listen to the criticisms. One NADAM member responded to Newman, "After every war, we seem to have a tendency to be supercritical and be anxious to say what is wrong with America instead of what is right with America. But I think it is healthy for NADAM to have criticisms, and I think we are big enough to profit by them."[10]

In short, the Association was confronted with a variety of new needs and the ongoing dilemma of how to respond to changes on individual campuses. The membership of men who had been drawn to the NADAM over the years had been, in fact, deans of men. Now, however, NADAM was being asked to include a broader membership including assistant deans, deans of students, deans of student personnel services, and other, new occupational titles. Newer members were coming from graduate training programs that many of the older deans of men had shunned only a decade earlier; many of the new members were younger men with backgrounds and education in psychology, sociology, as well as education and similar studies. The competition with the ACPA was heating up as well.

J.H. Newman further explored some of these pressing issues in his final comments to the membership. He reflected on the distribution of the membership and shared the types of colleges represented as well. He indicated, "The Association now has 170 members. This may be broken down as follows:

A&M Colleges 2
Municipal Colleges 9
State Universities 44
Religious backed L.A.S. Colleges ... 30
State Colleges 18
Private Universities 33
Technical Institutes 14
Religious Universities 5
Teachers Colleges 8
Private Colleges .... 9[11]

Newman also indicated that the NADAM had members in 46 of the 48 states as of 1949. He described the size of the institutions where the membership was located, ranging in size from fewer than 500 students to over 25,000 students. The bulk of the NADAM membership was clustered in those institutions ranging from 2500 to 5000 students (50 members) up through institutions with 10,000–20,000 students (28 members), a configuration that would resemble a bell curve with the smallest and largest institutions at either end of the curve.

Other business was conducted at the 1949 NADAM meeting and various speakers presented papers to the Association. Toward the end of the second day, Friday, April 15, the assembled deans heard the report of a committee created the year before and given the unwieldy name, Committee to Re-state the Aims of the Association. Donald DuShane of the University of Oregon was chairman. The committee had met with the NADAM executive committee and was now prepared to present its findings to the membership.

The first recommendation was a change in the wording of Article II of the Association. The article read, "The purpose of the Association is to correlate and study the most effective methods of service in the field of student welfare for men." The committee recommended that the article be changed, based on discussions at the Dallas 1948 meeting, especially the words, "men," "welfare," and "service." The revised article would read, " The purpose of the Association is to discuss and study the most effective methods of aiding students in their intellectual, social, moral, and personal development." The motion was passed by the membership.[12]

Other changes to articles were presented as well. Many of these items were discussed in Dallas, TX, the year before. A survey of the membership in 1947 and reported in 1948 precipitated many of the changes based on responses from the members. In part, it was acknowledged that many of the deans were no longer deans of men only. Many had become "deans of students" and others held related offices. One of the significant issues became the name of the Association as "deans and advisers to men" no longer seemed to reflect either the nature of the work nor the offices held by the members. In fact, it was reported that in 1948, only 70 men still held the title, "dean of men."[13]

A New Name

The final recommendation from DuShane referred to Article 1, the name of the Association. As DuShane reported,

> The question of a name change has been under consideration for many months. In the year prior to the Dallas meeting, a special committee

of the Association sent questionnaires, elaborate questionnaires, to all of the members on this point, and on other points. The matter was discussed at Dallas rather fully. It was voted to keep the present name. It was evident, however, that there was widespread feeling on the part of our members that the present name was not adequately descriptive, nor wholly accurate in terms of the membership of the Association, as it has developed in recent years;...[14]

DuShane indicated that although the sentiment in Dallas was that the name should be changed, there was no consensus on what that change might be. The final result was that the membership voted 60 to 11 not to change the name of the NADAM. Over the year between the meeting in Dallas and the meeting in Highland Park, IL, the committee had considered several options. DuShane indicated some of their deliberations and the preferences of the membership. "That the maleness or designation of maleness ought to be retained ....(laughter)...In our Committee's deliberations, we took that into consideration. We also tried to avoid one word which is a red flag to at least some of our members —"personnel."[15]

In the end, DuShane proposed the one name for the Association which, he claimed, had been acceptable both to his committee and to the Executive Committee, the National Association of Deans of Men and Deans of Students. Following this suggestion and the subsequent motion, there was considerable discussion of the name. One concern was that female "deans of students' now existed. Was the Association going to open its membership to women? In response, it was very clear that women were not completely excluded but it was not the intent of the deans to invite them either. In fact, it was obvious that there was significant antipathy for specific deans of women but no names were mentioned.

The debate continued with some, such as Robert Strozier of the University of Chicago, urging that the current name, NADAM, be retained. Others made other suggestions that rearranged the titles included in the name change, such as The National Association of Administrators of Student Life for Men. Others suggested that men with the title "dean of students" would come to the NADAM meetings anyway. Arno Nowotny of the University of Texas, who carried the title, Dean of Student Life, expressed his opinion that "I am always going to belong to NADAM no matter what you call it, and if some gal is smart enough to get elected Dean of Students in some institution and merit that title, I may let the old gal in. I don't know (laughter)."[16]

At the close of the conversation, the membership could not agree on a name change. The final vote was 65 to 36 to change the name to Dean of Men and Deans of Students but the vote tally was two votes shy of a 2/3 majority required by the by-laws to pass. So the membership voted instead to take yet another year to consider their choices and options and postpone a name change until the next annual meeting in 1950. Newman, who had stepped out of the role of chair briefly to urge passage of the new name, was clearly disappointed but conceded the battle. It was clear that sentiment for the past and the concept of the dean of men, even more so than the name, was not an easy symbol to give up. And the name change represented just that to many of the members—the loss of an important symbol, the dean of men.

When the deans reconvened in 1950, in Williamsburg, Virginia, (hosted by William and Mary University with 210 members present), the issue of the change in name was not addressed in a significant way. But in 1951, when the Association met in St. Louis, Missouri (hosted by Washington University and the Principia College in Illinois), the issue was once more at hand.

Association President Wesley Lloyd of Brigham Young University took advantage of his role as president to address the 222 members in attendance early in the conference. He told the group that this year's conference would not focus on campus issues and problems as usual, but would address two issues: the national emergency (by which he referred to reconstruction of American society after the war) and the role and position of the NADAM as it served personnel administrators across the country.

To address the NADAM issue, Lloyd reflected on his own involvement. "In 1938 when, as a new dean, I journeyed to Madison, Wisconsin, to attend the annual meeting and to get answers to the problems that crowded my inexperience.... I heard rumors that NADAM was a closed association dominated by a small circle of the older members. Yet, looking around for the first time, I found no closed brotherhood, but rather the greatest friendliness and welcome by all members regardless of length of membership."[17]

Lloyd further expressed his confidence in the Association to rise to address new challenges and issues. In particular, he asked the assembled deans, "Have we as an association delayed opportunities for national leadership among administrators in our field through being either too satisfied, ultra conservative, overly cautious, unduly aware of our chronological prestige, or simply disinterested in significant developments outside our own organization?"[18] By his comments,

Lloyd was referring primarily to the recent efforts to coordinate the student personnel services movement across the several national associations including the ACPA and the NADW representing the deans of women as well as representatives from the ACE and in particular, the Student Personnel Committee of the ACE. In these coordination discussions, the NADAM had been noticeably absent.

Lloyd then reiterated the history of the deans of men from the first 1919 meeting to the 1937 session where W. H. Cowley predicted the "disappearing deans of men." By 1950, Lloyd noted, "more than half of the institutions represented in NADAM had reorganized their student personnel work and appointed administrative officers with responsibilities for the broad areas of student life. The natural result was for many of the Deans of Men to be appointed the new offices. In a recent poll of men attending our meetings, it was found that fewer than half of the members held the title of Dean of Men."[19]

Lloyd continued to lay out the key issues for the NADAM. Did the Association wish to play a leadership role among student personnel associations or was it content to address only the needs of Deans of Men and Deans of Students? Further, it seemed that the votes over the past few years had reflected a preference for keeping the name, NADAM, and insisting that only men be able to attend the meetings. "It seems possible and advisable to retain these two provisions if we desire them regardless of the possible changes in the official name," Lloyd declared. Finally, he asked, "Are we paying too dearly for the luxury of isolation from other national organizations that are doing work closely related to our work?"[20]

Lloyd then introduced a group of deans he had asked before the meeting to openly discuss his remarks and the current sentiments of the Association. He indicated that they had already seen his speech in preparation. The deans who assembled in the front of the room were very familiar to the members. They were D.H. Gardner of the University of Akron; Robert Strozier of the University of Chicago; Robert Bates of Virginia Polytechnic Institute; Blair Knapp of Temple University, Dean Newhouse of Case Institute of Technology; and Victor Spathelf of Wayne University who chaired the session. (It should be noted that none of these men held the title "Dean of Men." Strozier, Newhouse, Spathelf, and Gardner were all Dean of Students. Bates was Director of Student Affairs, and Knapp was Vice President and Dean of Students.)

These six men carried on a lively and interesting discussion of the issues that Lloyd had laid out in his introductory remarks. To a large degree, the men laid out the existing connections between the NADAM and other groups. For example, Newhouse noted that Blair Knapp had on the previous Saturday,[21] "presided at the banquet of the American College Personnel Association in Chicago."[22] Newhouse noted that the Student Personnel Committee of the American Council on Education had also met on Saturday in Chicago and was in the process of asking all the organizations in student personnel to list the qualifications for the work they do.

Knapp noted that he was eager for the NADAM to continue to play a key role in the future of the student personnel area nationally. Knapp commented that he saw three roles in the student personnel movement—the technicians, the professors, and the administrators. By technicians, he referred to actual personnel administrators who administered tests and kept the records. The professors taught in the graduate programs. It was clear that the administrators were the deans and others like them who Knapp declared was the group that had "been less vocal" than the other two.

The NADAM, he stated, "has been an Association of personnel administrators and that our problems and interests are more in that direction than in any other." At the ACPA banquet, Knapp noted that at the head table there were three members of NADAM and that 'at the meetings which they had attended there were, conservatively, forty members of NADAM. I said, as far as I was concerned, I was not interested in a conversation about unification, preferring it come on the personal level."[23] In short, the six assembled deans argued that the NADAM needed to step into the leadership role that had been created for them.

In fact, many of the NADAM members had already been promoted or moved into positions that gave them the responsibility for not only student personnel functions on their campuses but also titles that clearly indicated that they were responsible for all students on their campuses, not just men. These changes had been occurring frequently over the several years after World War II ended. In the Secretarial report published at the end of the Proceedings of the NADAM, Fred Turner, finishing his fifth three-year term as Secretary, began posting, along with retirements and deaths of member deans, a list of Appointments and Promotions for each year. In 1951, he noted:

From Dean to President: 3
From Dean to Vice President: 4
From Dean to Associate to President: 1
New Deans appointed: 15
Assistant or Associate to Full Deanship: 9[24]

These new appointments and promotions along with the steady increase in membership indicated to the NADAM that they were on the rise in terms of national standing and needed to maintain that momentum.

To do so, they were coming to the consensus that the NADAM needed to "evolve," as some of the deans put it, and reassess and if necessary, rename the Association to reflect their broader aims to keep from painting themselves into a corner by only reflecting their past duties as deans of men. To win the competition between the ACPA and even the deans of women, change was necessary. Dean Arden French of Louisiana State University noted, "I have heard college presidents say that they do not care to get in the middle of a friction between the ACPA and the NADAM, and particularly the NADAW (laughter)."[25] Few things moved the NADAM toward action faster than the thought that the deans of women might advance ahead of them.

Later at the conference, Dean E.G. Williamson of Minnesota gave a speech on "Administration of Student Personnel Programs." Williamson also pushed the NADAM members in the direction of a more inclusive approach to the student personnel movement. Much of his speech was based on his forthcoming book, *Administration of Student Personnel Services.* Williamson acknowledged that he had not attended a NADAM meeting for ten years, much of his energy had been spent in work with the ACE on student personnel issues. But he quickly was able to tie himself to the history of the NADAM by noting that he had been an undergraduate student at Illinois while Dean Clark was Dean of Men.

Williamson noted that from his perspective, Dean Clark would not have found the new personnel methods to be "a roadblock for new developments of which he was unaware in his day because I saw him operate with new problems in the post-war period—with new adjustment problems, with a new type of student that he had not had prior to the war. I saw him make his adjustments first hand. In fact, I was one of his guinea pigs back in 1921."[26]

Williams went on to reassure the deans,

> It seems to me that the modern technology of student personnel work does not need to lose the individual touch which has been important in the history of the founding of this organization. In my twenty-five years at Minnesota, I have worked both sides of the street. I have been both a technician and at the same time, I have had to develop an administrative skill, which I did not learn in the classroom....

> That leads me to my last point that I want to make here. It has been a source of great embarrassment and great perplexity to me as an individual that the continued absence of official representation of this group [NADAM] at the other personnel meetings has been a cause of wonderment regarding our ideas and attitudes, and as to whether or not we are frozen in our development, or whether we wish to forge new methods of dealing with new problems. I would hope that whatever the merits of unification would be—and I have some reservations about it—that we would be able to establish a better liaison with the other personnel workers so that we can win their acceptance of our leadership.[27]

Williamson carried on in his discussion to address other issues in student personnel as reconfigured from business and industry to the college campus. He made note of the need for various adaptations for students and the need for creativity in matching techniques to varying situations. But the crux of his presentation was that the NADAM was missing in action when it came to the personnel movement and that condition was unacceptable.

On the second day, at the business portion of the conference, President Lloyd called the meeting to order and began to call for reports from the various committees. He quickly moved to call Vice President Spathelf to the table to introduce the work of the committee appointed to consider the future of the Association. The committee consisted of the six deans who had spoken at the beginning of the conference at Lloyd's invitation.

Spathelf addressed the membership and declared, "What the committee has thus presumed to do is to bend its efforts toward the clarification of our purposes and structure in order that we may more effectively address our concern to the role of leadership in meeting the needs of students. Indeed, to assume such a leadership role, we must say to ourselves: What kind of an organization are we; what our fundamental concerns are; and suggest the vehicles for attaining that which we feel are our mandatory obligations."[28] Spathelf described four commissions the committee recommended be created, one on professional relationships, another on principles and professional ethics, a third to address the development and training of student personnel administrators, and finally, a commission on program and practices evaluation. A motion was made and passed to establish the four commissions.

Then Spathelf introduced a motion to change Section II of the NADAM constitution. It had read: The purpose of the Association is to discuss and study the most effective methods of aiding students in their intellectual, social, moral, and personal development." The committee recommended the statement be amended to read:

> The institutions which are the constituent members of this Association are represented by those who are primarily concerned with the administration of student personnel programs in colleges and universities of the United States. Recognizing that many specialized abilities contribute to meeting student needs, this association seeks to provide and stimulate for the effective combination and utilization of all of these resources.
>
> As the student personnel program is affected by and affects the entire educational endeavor, this association cooperates with those agencies and associations which represent higher education, government, community resources, and specialized interests in student personnel work.[29]

Having gone this far in amending the purpose and aims of the Association, it seemed logical to address the next issue. Spathelf pointed out that there were some 36 different job titles in use among the members present. These titles ranged from Dean of Men to Adviser of Men, to Dean of the College, to Dean of Students. He made the point that the name of an association attempting to represent all of these occupational titles could not be based on a single title, such as Dean of Men. Therefore it was the recommendation of the committee that, "we designate this organization by general, functional description of that which describes our work. It is the recommendation of your committee that the name of this association be changed to the National Association of Student Personnel Administrators."[30]

The motion was seconded and then passed, almost unanimously, much to the delight of President Wesley Lloyd. He commented, "I need not indicate the trite possibility that this hour is a monumental one in the history of this Association, . . ."[31]

So after several years of discussion and debate, the Association that had begun in 1919 with a small group of six men in Madison, Wisconsin, changed its mission and its name.

The deans of men had labored long and hard to resolve their disagreements over their purpose and role. Over time, the NADAM had served a valuable purpose and been an organization that symbolized the spirit of the early deans of men for many of the members, both old and new. The greatest fear was that the informality and personal relationships that were hallmarks of the NADAM would be lost if the purpose and especially the name of the organization were changed. In some respects, the discussion was seen as the death of the old organization and all it represented to many of the members. On the other hand, it became very clear that as W. H. Cowley had suggested in 1937, the deans of men were going to disappear. The membership of the NADAM attempted to ensure that the organization they all valued, some might say loved, would not disappear as well.

The name change also signaled to the deans that their work and their philosophy toward that work had evolved. The compartmentalization of campus cultures into male and female was greatly diminished from what it had been in the early part of the twentieth century; however, the acceptance of women on campus was far from complete. Nonetheless, as the men of the NADAM and the women of the NADW could both acknowledge, change was inevitable. Now the question remained—what would the ongoing evolution of the student personnel movement look like and how would it affect campus cultures on colleges and universities across the United States? In the early 1950s, there was much that remained unknown. For the deans of men, now often deans of students or with similar titles, it meant more responsibility, more tasks, and many more students.

The new members of the NADAM, after March 31, 1951, known as NASPA, were given green ribbons to wear on their name tags to help introduce them to the current membership. As noted earlier, many of the new members were the men who would carry NASPA into the second half of the twentieth century. Many of the first deans of men had been professors from the arts and sciences, but the newer members of the Association were more inclined to have earned graduate degrees in the social sciences, such as sociology and psychology. Some, following Fred Turner's pattern, had earned advanced degrees in education or its variants, like educational psychology, educational administration, or more often, the administration of higher education. Regardless of their academic backgrounds and institutional affiliations, the men of the newly created NASPA would be challenged by each other and those outside the association. But at the least, they now had a common cause and a common organization.

9

A Retrospective Epilogue

Studying the deans of men from their inception in the early twentieth century to the 1950s presents several challenges. To borrow a phrase, the deans of men are "a riddle within an enigma," in that they defy easy definition. As Irma Voight, a dean of women and president of the National Association of Deans of Women in 1936 noted, the deans of men owed their existence to the deans of women. Had the deans of women not been so successful, the deans of men may not have been created at all.[1]

On the other hand, as colleges and universities grew in the post Civil War years, the complexity of the institutions became too much for a single president and a few faculty to manage. Though the ideal in the nineteenth century was "Mark Hopkins on one end of a log and the student on the other," the colleges and especially, the universities that began to emerge in the 1890s and beyond were more like corporations than seminaries. As president of Harvard, Charles Eliot bears a significant amount of responsibility for the creation of deans of men. Over-burdened with too many tasks and not enough time, Eliot named LeBaron Russell Briggs Dean of the College and responsible for issues related to students in 1890.[2]

Once Harvard did something, it was often widely copied by other colleges. Deans of the College were soon appointed at other schools, especially in the private colleges and universities in the Northeast. Even at the University of Illinois, where Thomas Arkle Clark became the first "official" dean of men in 1909, he was first given the title Dean for Undergraduates, in effect, a Dean of the College. Because in most cases there were no women students, the Deans of the College were ostensibly Deans of Men.

As the trend toward co-education spread beyond the experiment at Oberlin, to the wholesale commitment at the Universities of Iowa, Michigan, Minnesota, Chicago, and other institutions, the need for a faculty mentor (or matron) to represent and oversee the women became a necessity. As has been noted earlier and in other scholarship on the deans of women, it was often the dean of women who was appointed first, followed by a dean of men later. College and university presidents, as men, did not want to be the point person for issues or controversies related to female students. So they quickly appointed deans of women to deflect any problems or concerns. As Irma Voight argued, with some irony and her tongue firmly in her cheek, the deans of men were, after all, an afterthought.[3]

However, in reality, deans of men, like the deans of the college, were a good idea. By appointing a man with the proper character and right demeanor to manage the student body or at least the male students, and to address all the various and sundry issues that might come up regarding their welfare, from discipline to housing to illness to any out-of-class activities, was sheer genius. The best approach to sound management was delegation and most college and university presidents were eager to delegate the day-to-day supervision of students to someone else.

On the coeducational campus, however, deans of men and deans of women were not equals. By definition, women on the coeducational college campus were the "other," the "un-men," and remained a minority populations that warranted special consideration and attention but were never the same as men. Deans of women were advocates for women students. But by default, they were also symbols of the female presence on campus for all women, including female faculty and staff. This symbolic aspect of the dean's position also meant that the dean of women was an easy target for both praise and criticism.

Deans of men, on the other hand, were faculty members who were charged with the responsibility for a myriad of duties, from maintaining social order to individual advisement and counsel to any activity that did not clearly fit into the academic realm. Once a male student walked into a classroom, he was the responsibility of the professor. As soon as he left the classroom, he was the responsibility of the dean of men. Even absences from class were to be regulated by the dean, who reviewed and pronounced a missing student "excused" or "not excused."

To the teaching faculty, like the presidents, these onerous tasks were tedious and a distraction and they were often only too glad to have a dean of men to address them. As Laurence Veysey has described the "gulf between students and faculty" in the early twentieth century, the two groups were on different tracks. Young men were in pursuit of a life unbounded by serious study and scholarship whereas the faculty were often myopic in their attention to their discipline or field of study, often to the exclusion of any extracurricular activities or events. As Veysey noted, the slogan often found in student rooms of the time was "Don't let your grades interfere with your studies."[4] The goal of a college education was to live a life away from the control of parents and others in the company of others intent on a good time and various entertainments. A "gentleman's C" was considered an appropriate accomplishment in light of the many other options available for study. It was a classic tale of the class of two cultures, students and college faculty.

Into this mixed bag of intentions and expectations between students and faculty came the dean of men. As most students were male, the deans were not public symbols of their sex as were the deans of women. They could stride across the campus secure in their right to assume a certain level of leadership and responsibility. But in their new roles, they struggled for a clear definition of their role and function. As Stanley Coulter of Purdue exclaimed at a NADAM meeting, "When I wrote to the Board to ask them what were the duties of the dean of men, they wrote back and told me they didn't know but that when I found out, to let them know."[5]

On the smaller, private liberal arts campuses, Deans of the College such as Briggs at Harvard, Hawkes at Columbia, Speight at Swarthmore, Christian Gauss at Princeton viewed their role as a mix of academic and co-curricular activities. On the public campuses, Deans of Men like Clark, Goodnight, Rienow, Coulter and others had a more specific role. Although most of the deans of men continued to teach in their discipline even after becoming deans, the men at the larger schools were often encumbered with too many administrative tasks to be able to maintain a regular teaching load. By taking on a wide range of tasks, the deans became the omnipresent face of the ultimate college authority to students.

By 1931, the Office of Dean of Men at Illinois had 352 visitors per day. Many students sought loans or work to offset the loss of funds due to the Depression.[6] Scott Goodnight reported that he personally interviewed an average of 19 students per day in his office.[7] Any student with a problem or need of any kind would seek out the dean. On larger campuses, deans of men became further removed from the academic arena as their positions evolved. This process of decoupling the dean of men from the faculty left some deans in a position of limbo between the president and provost or other chief academic officers and the faculty. Their best recourse in attempting to define their roles and sort out their workloads was to be found in talking to other men in similar positions on other campuses. As a result, the annual conferences of the NADAM became an essential and welcome respite as is clearly reflected in the proceedings of those meetings.

The deans of men, early in their evolution, often referred to their position as a "personality." By invoking this explanation, they defined the position as an extension of the individual dean. In many respects, the description is appropriate but not very clear. A better explanation is that often, the man selected to be a dean of men was a person, almost always a current faculty member, who had a natural affinity for working with young men. A characteristic that the early deans shared was a compassionate commitment to working with students. From Briggs at Harvard to Clark at Illinois to Gauss at Princeton and beyond, these men enjoyed the role of advisor, counselor, and mentor to young men. The activities of the dean's office brought them great satisfaction and they valued the work. They could often see the benefits of their advice as young men changed their behavior or altered their course, or simply left the dean's office with a problem solved.

The problems or challenges young men faced on the college campus ranged from lack of funds to homesickness to conflicts with faculty, fellow students, or even absent parents. Violations of campus rules and regulations were a constant source of challenge for the deans of men as well. Though few deans relished the role of campus disciplinarian, as deans they were charged with the control and direction of the student body. They even found the process of discipline an opportunity to have a conversation with a young man ("boy") about other problems or missing elements in his life on campus. The worst cases required expulsion from the college or university. But often enough, even an expelled young man was allowed to re-enroll after a period of absence, a time-honored tradition that can be traced back to Harvard in the 1700s when young men were tossed out for card playing, drunkenness, or contracting "the itch" from local prostitutes. After a term away and a period of penitence (and often appropriate compensation by parents), a young man could re-enroll and continue his studies.[8] Contemporary higher education institutions still practice the "go away and then return" process for serious violations of campus regulations in all but the most egregious cases.

The challenges of being the dean of men were significant. Often, the dean acted on his own judgment, making decisions based on moral standards thought to reflect the best interests of the college. On many campuses, a special committee of faculty, as existed at Wisconsin and even Illinois, was put in place to review campus behavior. But it was typically the dean of men who administered the enforcement of campus rules. These practices made the dean an easy target for the wrath and ridicule of students, many of whom viewed the dean of men as the personification of the parents they left at home. As parent substitutes, deans could be reviled when they corrected bad behavior or embraced when they provided support and comfort. In either case, it was a very personal as well as professional role for the dean of men.

When major issues affecting students emerged on campus, it was often the dean of men who was to interpret or implement campus policies. As young men of color sought enrollment, it was the dean of men who bore the responsibility for their oversight or exclusion. Scott Goodnight reflected on the "Negro" student population in Madison in a letter, noting that one young man would be better served by seeking a more appropriate venue for his education, somewhere other than the University of Wisconsin. On other campuses, especially in the Northeast, men of Jewish heritage as well as African Americans were not admitted. It was left to deans, such as Herbert Hawkes or Christian Gauss, to spread the word. Similarly, the role of women on campus was a touchy subject. Many of the deans were tolerant but not eager to address the issues of women students and the resentments were often expressed in sarcasm or humor toward women and especially, the NADAW.

Not surprisingly, on many days, the physical and emotional challenges of the office were exhausting. To persist, the work of the dean of men had to be ingrained in the officeholder. The work of the dean had to become a "labor of love," a calling, as it were, and this was in fact a sentiment expressed often by many of the early deans. At the root of their argument against graduate training as preparation for the dean's job as insufficient was the idea that a man had to have the "personality" to be a dean or else there was no point in trying to become one. Graduate courses were ineffective if one did not have the temperament to be a dean

In light of the demands of the office, the annual gatherings of the deans of men at the National Association of Deans and Advisers of Men were rare opportunities to sit among men of similar temperament, common goals, and kindred spirits and simply relax and take comfort that one was surrounded by others who were sympathetic to the challenges of the position. At the NADAM meeting, there was common interest and support for the work of the dean. These meetings provided a respite from a constant flow of students, concerned parents, and indifferent faculty. The deans could wrestle with issues that other deans had also confronted or even solved. So it is not surprising that a primary concern of the NADAM membership in the late 1940s was less the name change under consideration but instead the threat that a name change might bring an end to the camaraderie and informality that many of the deans had enjoyed for quite some time.

In many respects, the challenge to the NADAM from the new and often younger deans or advisers in the late 1940s—that the NADAM was more of a social club that was run by a small, elite group of deans—was true. It was a club of fairly exclusive members. The membership of the NADAM began to climb significantly in the late 1930s and 1940s. But even at that, the NADAM annual meeting first registered 200 or more participants in 1949.[9] When the name of the NADAM was changed to NASPA in 1951, the exclusivity of the organization was altered but the same sentiments persisted. The role of dean of men would evolve as the titles of the deans changed to deans of students, and eventually vice presidents for student services.

But the intimate nature of the relationships between the men continued over time and the "club" atmosphere persisted long after 1951. For competing associations, such as ACPA and National Association of Women Deans, Administrators, and Counselors (NAWDAC), the nature of the "old White guys" club that was NASPA was often used against it. Women and younger professionals often declared that they felt more comfortable in the "other" associations. The approbation of "club" continued to characterize NASPA for years after the NADAM had ceased to exist and began to break down only when women were accepted for membership and eventually began to assume leadership roles in the late twentieth century.

The evolution of the role of the dean of men and the eventual absorption of the positions of deans of men and women into the student personnel movement can be seen from an academic as well as social perspective. The third meeting of deans of men held at the University of Iowa in 1921 was attended by an English professor, two German professors, a professor of mechanical engineering, three professors of physics, two professors of chemistry, a biology professor, a professor of romance languages, a professor of mathematics, a professor of geology, and three men who listed no faculty affiliation. But all sixteen men were deans of men. As the student personnel movement advanced in the 1920s, many more men who had academic roots in psychology, sociology, and education entered the field of dean of men. So the emphasis gradually changed from an arts and sciences perspective to a social sciences perspective.

The student personnel movement had it roots and early beginnings in the application of psychology to business and industry and to some extent, the military. Walter Dill Scott, who was a prime contributor to the use of personnel psychology, used his knowledge of psychology to study advertising and sales early in his career. When World War I broke out, he offered his skills to the Army for classification of officers and enlisted men.[10] Scott and others began to apply the personnel psychology to college students after World War I had ended. The movement was an extension of the social efficiency efforts that grew out of the Progressive era. Through activities and applications of vocational guidance, time and motion studies, and the application of scientific principles to human endeavor, the student personnel movement sought greater efficiencies in higher education by focusing on the skills, aptitudes, and competencies of college and high school students.

These efforts led to greater use of tests and measurements as well as personal interviews, record keeping, and vocational counseling. Some of these efforts had first been utilized by the Young Men's Christian Association (YMCA) in the early twentieth century, as the organization tried to help young men in large metropolitan areas find employment. Gradually young men, occupational counselling, and personnel psychology all intersected in colleges and universities. But to the early deans of men, many of whom were trained in the liberal arts, not in psychology or related areas, the personnel movement seemed artificial and foreign. In fact, many of the early deans referred to the personnel movement practitioners as "scientific nuts."

Gradually, the benefits and accumulated knowledge of and about students and their adjustments in college gained through the personnel movement became hard to dismiss. In addition, the full court press applied by the American Council on Education and other agencies to expand the personnel movement proved to be unstoppable. The earliest deans of men, the chemists, biologists, physicists, and the like, had to concede ground to the new social sciences and their ability to understand and interpret human behavior.

At the same time, the basic idea of sitting with a young man one on one, as Clark, Coulter, Rienow, and Briggs had done for years, and learning about his fears, expectations, and need for the advice of a mentor/ role model/father figure, could be lost in the rush to measure, assess, and evaluate. The accumulation of data was a benefit to the institution, but the ability of getting to know the "whole person" as many deans had done over time, was a difficult task to maintain in the face of specialized use of interview forms, testing, and personnel assessments. The deans of men respectfully resisted these modern efficiencies in favor of the familiar getting to know a young man on a personal level.

However, when the enrollments on most college campuses grew exponentially in the post World War II era of the late 1940s and

1950s, the option to create routines and mechanisms designed to manage and control the vast numbers of students on campus became very attractive and much more efficient. Thomas Arkle Clark had hired a small group of assistants in the 1920s to manage his office at Illinois. Yet he only had 7500 male students to oversee. In the post-war era, those numbers grew rapidly and the ability to keep up became paramount. Soon, management and managerial systems became the norm on many campuses. Even the small schools found it difficult to maintain social order and normalcy, especially in the face of large numbers and mature, older students, as veterans flooded the campus.

Though the concept of deans of men was still useful and necessary, many became deans of students and soon assumed the responsibility for both male and female students on campus. The social demands were for traditional roles for men and women to be reflected on the campus. Deans of students were responsible for the management of larger numbers of students, and far fewer women, and overseeing new specialists in the areas of career counseling, residence halls, veterans affairs, and registration activities. Informal chats with students were pleasant but not a very efficient use of time. As the numbers increased, so did the bureaucratic nature of the college or university, and in turn, the job of the dean.

As time passed, and student veterans of the 1940s and 1950s moved on, the Baby Boom that followed World War II sent a new wave of students to college in the 1960s. The sons and daughters of a new middle class had high expectations for themselves and their educations. Once again, the numbers increased significantly. Many public institutions were awash in federal and state money, research opportunities tied to business, industry, the military, and above all, lots of students. As colleges exploded in size, the social issues of civil rights, the war in Vietnam, and related issues were logs dumped on a smoldering fire of unease, discontent, and impersonalization. As Roger Geiger has noted in his discussions of the universities in the twentieth century, there was a dissolution of the consensus that had held colleges and universities together.[11] Student disruptions and dissatisfaction were rampant in the 1960s in particular as campuses erupted over even the slightest friction and on many larger campuses, the great social issues of the day were debated in the streets.

Could some of the disruptions of the '60s have been prevented by a more intimate knowledge of students and student life? Did the reliance on measurement, testing, the bureaucracy of mass registrations, and the rallying cry of the sixties protests, "don't treat me like a number" create a less human or at least humane environment on many campuses? Possibly. But would the intimacy of the early deans of men have provided a cathartic solution? Possibly. The lessons from the past, the student-friendly dean who created a personal connection and commitment that many students felt and acknowledged was no longer present. In the past, students and alumni at Wisconsin, Illinois, Columbia, Harvard, Princeton, and other schools had been eager to stay in touch with or at least hear about Scott Goodnight long after he had left Madison, or to attend Tommy Arkle's funeral to pay respect, or to raise from their own pockets an annuity pension fund for L.B. Briggs. There was a connection and mutual respect and admiration shared between the deans and their students.

The role and functions of the deans of men were lost as institutions grew larger and more complex. The roles and functions that the deans performed, the connections and relationships with students, were re-created after the 1960s in other forms. When Thomas Jones wandered the campus of the University of South Carolina to talk to students, he was dismayed to find that many students did not know the history of the university nor did they have intimate relationships with their faculty or even many other students. He conscripted a young library science professor, John Gardner to develop a program to help orient new students to college. The idea became known as the Freshman Year Experience.[12] The need for students to experience the human touch personified by deans of men in the past has been documented in research from the 1970s through the present by Alexander Astin, who has characterized the value and benefit of student involvement in the modern era.[13] Students persist to graduation more frequently if they feel "connected" to the college or university through faculty interactions, out-of-class experiences, and programs designed to acquaint them with the campus. George Kuh has used the National Survey of Student Engagement to document the same needs for connections to the campus in his contemporary research.[14]

It is never possible to go backward in time, and it must be acknowledged that the deans of men are long gone. However, they did leave a legacy of commitment, compassion, and care that should be honored, respected, and appreciated. At their best, they worked hard to support and serve students in the best ways they knew how. Their legacy, it is hoped, may be perpetuated in some small way through this book.

Notes

1 The Rise and Demise of Deans of Men: A Historical Perspective

1. Burton Bledstein (1976). *The Culture of Professionalism: The middle class and the development of higher education in America*. New York: W.W. Norton.
2. John Thelin (2004). *The History of Higher Education in America*. Baltimore, MD: The Johns Hopkins Press.
3. Frederick Rudolph (1990). *The American College and University*. Athens, GA: The University of Georgia Press.
4. Laurence Veysey (1970). *The Emergence of the American University*, University of Chicago Press.
5. Frederick Rudolph (1990). *The American College and University*. Athens, GA: University of Georgia Press.
6. Lawrence Veysey (1965). *The Emergence of the American University*. University of Chicago Press.
7. R. Brown (1928). *Dean Briggs*. New York: Harper and Co.
8. Ibid.
9. Robert Schwartz The Rise and Demise of Deans of Men, *Review of Higher Education*, 26(2), 217–239.
10. Edward Saveth (1988). The Education of an Elite. *History of Education Quarterly*, Vol. 28, No. 3 (Autumn, 1988), pp. 367–386.
11. Ibid.
12. Thomas Arkle Clark (1921). *Discipline and the Derelict: Being a Series of Essays on Some of Those Who Tread the Green Carpet*. New York: MacMillan.
13. Stanley Coulter (1934) in *Secretarial Notes of the 16th Annual Conference of the National Association of Deans and Advisers of Men held at Northwestern University, Evanston, Illinois, 1934*. Lexington: University of Kentucky Press.
14. Robert Rienow in Secretarial *Notes on the National Association of Deans and Advisers of Men, 1928*, p. 28.
15. Scott Goodnight in *Secretarial Notes on the National Association of Deans and Advisers of Men, 1934*, p. 28.
16. Ibid., p.55.

17. Early History of the Association in *Secretarial Notes on the National Association of Deans and Advisers of Men, 1934,* pp. 27–56.
18. Ibid.
19. For more on the role of the "organized," see Helen Horowitz's book, *Campus Life* (1987), in which she carefully documents the direction of campus life by various student groups.
20. Stanley Coulter (1928) in *Secretarial Notes on the National Association of Deans of Men, 1928.*
21. Abraham Flexner's reviews of social work, colleges, and schools of medicine (especially those for African Americans) and other areas of higher education early in the twentieth century were notoriously difficult and often castigated the institutions under review. Flexner is now credited with major changes that improved or at least addressed these concerns. See his *Medical Education in the United States and Canada* (1910) and *The American College: A Criticism* (1908) as examples.
22. See Gertrude Martin (1927) Report of the Committee on the History of the NADW. In the *14th Yearbook of the National Association of Deans of Women*, pp. 212–213.
23. Stanley Coulter (1928) in *Secretarial Notes on the National Association of Deans of Men, 1928.*
24. Robert Rienow (1931) in *Secretarial Notes on the National Association of Deans of Men, 1931,* pp. 110–113.
25. For a more extensive discussion on the student culture of the 1920s, see Paula Fass *The damned and the beautiful: American youth in the 1920's* (New York: Oxford University Press, 1977).
26. Patricia Graham (1978) Expansion and Exclusion: A history of women in American higher education. *Signs: Journal of Culture and Society, 3,* pp. 759–773.
27. William Alderman in *Secretarial Notes on the National Association of Deans of Men.*
28. L. Blayney (1928) College deans and the report of the Carnegie Commission. In *Secretarial notes of the 10th annual meeting of the National Association of Deans of Men*, pp. 21–30.
29. Ibid.
30. Ibid.
31. Ibid.
32. Bursely, *Notes on the National Association of Deans of Men*
33. Ibid.
34. See R. T. Von Mayrhauser (1999). Walter Dill Scott. *American National Biography* (Vol. 19, pp. 507–510). New York: Oxford University Press.
35. Esther Lloyd-Jones and L. B. Hopkins both wrote books based on their Northwestern experiences under Scott.
36. Robert Schwartz (2000, Fall). F.F. Bradshaw, A Student Affairs Pioneer, *Southern Association for College Student Affairs Journal.*
37. Francis F. Bradshaw, The Dean of Men's Preparation for His Work. *Secretarial Notes on the National Association of Deans of Men, 1931*, pp. 104–108.

38. Ibid.
39. Fred Turner, Report of the Committee on Preparation for Work as a Dean of Men. *Secretarial Notes of the National Association of Deans and Advisers of Men, 1936*, pp. 16–50.
40. Ibid.
41. Ibid.
42. Harold Speight in *Secretarial Notes of the National Association of Deans and Advisers of Men, 1936*, p. 49.
43. F.M. Massey in *Secretarial Notes of the National Association of Deans and Advisers of Men, 1936*, pp. 50–51.
44. Kathryn Moore (1976). Freedom and Constraint in Eighteenth Century Harvard, *The Journal of Higher Education*, Vol. 47, No. 6, pp. 649–659; Jennings Wagoner (1986). Honor and Dishonor at Mr. Jefferson's University, *History of Education Quarterly*, Vol. 26, No. 2, pp. 155–179.

2 THE PIONEER: THOMAS ARKLE CLARK, DEAN OF DEANS

1. "Tommy Arkle" Obituary, *Time* magazine, August 31, 1931.
2. T. A. Clark, *Secretarial Notes of the 13th Annual Meeting of the National Association of Deans of Men, 1931* (Lawrence, Kan.) p. 101.
3. T. A. Clark, *Secretarial Notes of the 13th Annual Meeting of the National Association of Deans of Men, 1931* (Lawrence, Kan.) p.101.
4. T. A. Clark, *Secretarial Notes of the 13th Annual Meeting of the National Association of Deans of Men, 1931* (Lawrence, Kan.) p. 101.
5. T. A. Clark papers, University of Illinois Student Life Collections, University Archives, 41/2/20, Box 1.
6. T. A. Clark papers, University of Illinois Student Life Collections, University Archives, 41/2/20, Box 1.
7. T. A. Clark, *Secretarial Notes of the 13th Annual Meeting of the National Association of Deans of Men, 1931* (Lawrence, Kan.) p. 102.
8. The Career of Thomas Arkle Clark, T. A. Clark papers, University of Illinois Student Life Collections, University Archives, 41/2/20, Box 1.
9. Jana Nidiffer & Timothy Reese Cain (2005). "Elder Brothers of the University: Early Vice Presidents in Late Nineteenth-Century Universities," *History of Education Quarterly*, 44, (Winter), 487–523.
10. Nidiffer and Cain, "Elder Brothers," p. 519.
11. "Origin and Development of the Office of Dean of Men in the University of Illinois" (no author) Clark papers, Box 1, University of Illinois Student Life Collections.

    [While no author is listed for this document, it would appear to have been written by someone with intimate knowledge of the circumstances over a 15-year period from 1900 to 1915. The author may have been Clark or Clark dictating to an assistant such as Arthur Warnock,

Assistant Dean of Men, 1910–1921 or perhaps Fred Turner, Assistant Dean of Men, 1922–31.]

12. "Origin and Development of the Office of Dean of Men in the University of Illinois" (no author). Clark papers, 41/2/20, Box 1, UISCL.
13. Inflation Conversion Factors for Dollars 1665 to Estimated 2015, Robert Sahr, www.oregonstate.edu/dept/pol_sci/fac/sahr (Retrieved August 30, 2005).
14. "Origin and Development of the Office of Dean of Men in the University of Illinois" (no author). Clark papers, 41/2/20, Box 1, UISCL.
15. Clark papers,.41/2/20, Box 1, UISLC.
16. Clark papers,.41/2/20, Box 1, UISLC.
17. Clark papers, UISLC, 41/2/20, Box 1, UISLC.
18. Clark papers, UISLC, 41/2/20, Box 1, UISLC.
19. Clark wrote a chapter titled, "If I Were Adviser to Girls" for a book by Jameson, K.W. & Lockwood, C.R. (Eds.) (1925). *The Freshman Girl* (pp. 111–125). New York: D.C. Heath and Company.
20. Clark papers, UISLC, 41/2/20, Box 1, UISLC.
21. Clark papers, UISLC, 41/2/20, Box 1, UISLC.
22. Clark papers, UISLC, 41/2/20, Box 1, UISLC.
23. Helen Horowitz, (1988). *Campus Culture: Undergraduate Cultures from the Late 18th century to the Present.* Chicago: University of Chicago Press.
24. Quaker Oats advertisement, *Saturday Evening Post* magazine. Clark papers, Box 1, U University of Illinois Student Life Collections.
25. Clark papers, UISLC, 41/2/20, Box 1, UISLC.
26. Clark papers, UISLC, 41/2/20, Box 1, UISLC.
27. Clark papers, UISLC, 41/2/20, Box 1, UISLC.
28. DiMartini, J. R. (1974) "Student Culture as Social Change Agent," *Journal of Social History*, 9, pp. 526–541
29. Clark papers, UISLC, 41/2/20, Box 1, UISLC.
30. Clark papers, UISLC, 41/2/20, Box 1, UISLC.
31. DiMartini, 1974.
32. Ralph Jones, *Proceedings of the National Education Association, 1910*, p. 556.
33. T. A. Clark, *Proceedings of the NEA*, 1910, p. 557.
34. F. Turner, *Banta's Greek Exchange,* October, 1931, p. 393.
35. Horowitz, 1988
36. *The Sunday Eight O'clock: Brief Sermons for the Undergraduate* Urbana, IL., 1916
37. *Facts for Freshmen, 1916*, Clark papers, UISLC, 41/2/20, Box 1, UISLC.
38. Letter dated November 11, 1930, Clark papers, 41/2/20, Box 1, UISLC.
39. Pinkerton Detective Agency -hired to protect trains from pilfering and damage.
40. Letter from Harold G. Baker, U.S. Attorney to E.C. Yellowley, Federal Prohibition Administrator, Chicago dated March 5, 1928. T. A. Clark

papers, University of Illinois Student Life Collections (UISLC), University Archives, 41/2/20, Box 2.

41. Letter from Harold G. Baker, U.S. Attorney to E.C. Yellowley, Federal Prohibition Administrator, Chicago dated March 16, 1928. T. A. Clark papers, UISLC,, 41/2/20, Box 2.
42. Eliot Ness became very well known later in his career for his work in law enforcement, most notably his determined war with Al Capone in Chicago. Ness headed up a team of Treasury agents who were nicknamed "the Untouchables" for their lack of corruption, unlike many others in law enforcement during the Prohibition period. Ness was elevated to a folk hero first in a television series in the 1960s called, *The Untouchables* and again in several movies, most recently in 1987, by the same name.

    When the "real" Eliot Ness traveled to work with Dean Clark in 1927, it was shortly after Ness had returned to the University of Chicago for a second degree in criminology studies. After Chicago, Ness went to work with the Treasury Department in Chicago. Ness then transferred to the Prohibition Bureau where he was one of the 300 men responsible for prosecuting the flourishing Chicago bootlegging industry, which led to his visit to Champaign-Urbana. www.crimelibrary.com/ness/nessmain retrieved August 30, 2005).
43. Interview with William O'Dell, Class of 1931, University of Illinois Student Life, 1928–1938, Oral History Project, interview transcript dated November 7, 2000, University of Illinois Student Life Collections.
44. "To the Parents of Undergraduates" dated November 1, 1924, Clark papers, 41/2/20, Box 1, University of Illinois Student Life Collections.
45. Interview with Royal Bartlett, Class of 1931, University of Illinois Student Life, 1928–1938, Oral History Project, interview transcript dated March 18, 2001, University of Illinois Student Life Collections.
46. Ibid.
47. T.A. Clark (1922) Discipline and the Derelict: Being a series of essays on some of those who tread the green carpet. New York: The Macmillan Company. p. 25.
48. *New York Times*, Obituaries, July 19, 1932, Thomas Arkle Clark.

3 THE PATERNALISTS

1. Secretarial Proceedings of the National Association of Deans of Men (NADM), 1934, p. 33.
2. Secretarial Proceedings of the NADM, 1934, p. 33. In fact, Strauss intended the minutes he took be seen only by Goodnight. Strauss had prepared a more sedate and official account of the meeting but Goodnight refused him and encouraged Strauss to publish the humorous version instead (letter to Goodnight from Strauss, November 4,

1919, and response from Goodnight, November 15, 1919, UWisconsin Archives, Student Affairs, 19/2/1-4, Envelope 1).

3. Secretarial Proceedings of the NADM, 1934, p. 33.
4. Summer sessions notes—Goodnight papers, UWisconsin Archives, 19/2/1-1.
5. Goodnight personnel records (handwritten), Goodnight papers, University of Wisconsin, Box 19/2/12.
6. *Milwaukee Journal*, May 18, 1945, p11- UW Archives, 19/2/3.
7. Ibid.
8. Goodnight memorial biography, Goodnight papers, Box 19/2/1.
9. In a newspaper interview published in 1965 on the occasion of his 90th birthday, Goodnight acknowledged that he "quit the German department in 1917 because "I didn't get along with the older members of the department staff." Russell B. Pyre, *Wisconsin State Journal*, January 2, 1965, "A Visit with Dean Goodnight at 90."
10. President's Report of 1896, cited in Thwaites, History of the University of Wisconsin, retrieved on-line 1/07/04 [http://www.library.wisc.edu/etext/WIReader/Thwaites/Chapter12.html#Section01].
11. Solomon, Barbara 1985. *In the Company of Educated Women: A History of Women and Higher Education in America*. New Haven: Yale University Press.
12. P. A. Graham (1978). Expansion and Exclusion: A History of Women in American Higher Education. *Signs: Journal of Women in Culture and Society*, 3, 759–773.
13. Radio address, circa 1922–23 school year. (No title). University of Wisconsin Archives, Student Affairs, Office of General Correspondence files (S.H. Goodnight) 1920–1945, 19/2/1-1, Box 11, Radio broadcasts, 1922–1923 (Wisconsin Educational Radio).
14. Goodnight, Secretarial Proceedings of the National Association of Deans of Men, 1934, p. 55.
15. Goodnight radio speech, circa 1922. (No title) University of Wisconsin Archives, Student Affairs, Office of General Correspondence files (S.H. Goodnight) 1920–1945, 19/2/1-1, Box 11, Radio broadcasts, 1922–1923 (Wisconsin Educational Radio).
16. Goodnight Correspondence files, 1920, H-J, Hal Hoag letter. UWA.
17. University of Wisconsin Archives, Student Affairs, Office of General Correspondence files Goodnight files, 19/2/3-1, Box 13—personnel cards.
18. University of Wisconsin Archives, Student Affairs, Office of General Correspondence files Goodnight files, 19/2/3-1, Box 13—personnel cards. Salary equivalencies taken from the U.S. Bureau of Labor Statistics inflation calculator, www.bls.gov, retrieved 1/13/04.
19. University of Wisconsin Archives, Student Affairs, Office of, General Correspondence files Goodnight files, 19/2/3-1, Box 13—personnel cards.
20. Retrieved on line, 1/7/04 from the University of Wisconsin University Communications Office at: http://www.news.wisc.edu/packages/chancellorsearch/pastchan.html

21. University of Wisconsin Archives, Student Affairs, Office of General Correspondence files (S.H. Goodnight) 1920–1945, 19/2/1-1, Box 1, Letters Received and Sent 1921–22.
22. University of Wisconsin Archives, Student Affairs, Office of General Correspondence files (S.H. Goodnight) 1920–1945, 19/2/1-1, Box 1, Letters Received and Sent 1921–22.
23. University of Wisconsin Archives, Student Affairs, Office of General Correspondence files (S.H. Goodnight) 1920–1945, 19/2/1-1, Box 2, Letters Received and Sent 1921–22.
24. Letters between George Banta and Goodnight, dated June, 1923, University of Wisconsin Archives, Student Affairs, Office of General Correspondence files (S.H. Goodnight) 1920–1945, 19/2/1-1, Box 2, Letters Received and Sent 1922–23.
25. Untitled report from the Sub-Committee on Student Living Conditions and Hygiene, circa 1943. University of Wisconsin Archives, Student Affairs, Office of General Correspondence files Goodnight files, 19/2/3–1.
26. Editorials from the *Daily Cardinal* newspaper, January 21–25, 1930. University of Wisconsin Archives, Student Affairs, Office of General Correspondence files (S.H. Goodnight) 1920–1945, 19/2/1-1, Box 2.
27. University of Wisconsin Archives, Student Affairs, Office of General Correspondence files (S.H. Goodnight) 1920–1945, 19/2/1-1, Box 2, Letters Received and Sent 1921–22.
28. *Wisconsin State Journal*, January 26, 1960—Goodnight Correspondence, Box 1, UWA.
29. Letter to President Frank on the Committee on Student Life and Interests, October 1, 1926, Goodnight correspondence file, UWA.
30. Radio address, 1922–1923 school year. University of Wisconsin Archives, Student Affairs, Office of General Correspondence files Goodnight files, 19/2/3-1.
31. Radio interview (1943). University of Wisconsin Archives, Student Affairs, Office of General Correspondence files Goodnight files, 19/2/3-1.
32. *Wisconsin State Journal* article, (n.d.) (1940). "Very Few 'Reds' on U.W. Campus, Goodnight Says. University of Wisconsin Archives, Student Affairs, Office of General Correspondence files Goodnight files, 19/2/3-1.
33. Ibid.
34. Excerpted from Duties of the Office of the Dean of Men, University of Wisconsin, 1931. University of Wisconsin Archives, Student Affairs, Office of General Correspondence files (S.H. Goodnight) 1920–1945, 19/2/1-1.
35. Ibid, p. 5.
36. Annual report of the Office of the Dean of Men, 1938, p. 21. University of Wisconsin Archives, Student Affairs, Office of General Correspondence files (S.H. Goodnight) 1920–1945, 19/2/1-1.

37. Ibid, pp. 5–6
38. Ibid, p. 12.
39. Report of the Office of Dean of Men, August, 1938, pp. 1–2. University of Wisconsin Archives, Student Affairs, Office of General Correspondence files (S.H. Goodnight) 1920–1945, 19/2/1-1.
40. Report of the Office of Dean of Men, August, 1938, p.36. University of Wisconsin Archives, Student Affairs, Office of General Correspondence files (S.H. Goodnight) 1920–1945, 19/2/1-1.
41. Goodnight letter dated January 16, 1960. University of Wisconsin Archives, Student Affairs, Office of General Correspondence files (S.H. Goodnight) 1920–1945, 19/2/1-1.
42. Interview with Russell Pyre, *Wisconsin State Journal*, January 2, 1965.
43. Ibid.
44. Floating university notes, 1928 University of Wisconsin Archives, Student Affairs, Office of General Correspondence files Goodnight files, 19/2/3-1.
45. "Dean With a Big Stick Defended Students Too." *Wisconsin State Journal*, January 17, 1960.

4 The Academics: Early Deans in the Liberal Arts Colleges

1. R.W. Brown (1926). *Dean Briggs*. (New York: Harper and Brothers.)
2. Ibid.
3. LeBaron Russell Briggs (1999). *American National Biography* New York: Oxford University Press. pp. 541–542.
4. Ibid.
5. Charles William Eliot (1908). *University administration.* Boston: Houghton Mifflin Co.
6. LBR Briggs (1901). *School, College, and Character* New York: Houghton Mifflin
7. It must be noted that Radcliffe was created in direct response to Charles Eliot's refusal to admit women to Harvard.
8. See Briggs' book, *Girls and Education* (1911)/ New York: Houghton Mifflin (later re-published in 1914 as *Letters to College Girls and Other Essays*.
9. A. R. Nelson (2001). *Education and Democracy: The Meaning of Alexander Meiklejohn, 1872–1964.* Madison, WI: University of Wisconsin Press.
10. Martha Mitchell (1993). Alexander Meiklejohn. In *Encyclopedia Brunoniana* Providence, RI: Brown University. Retrieved online, July 11, 2009 from www.brown.edu/Administration/News_Bureau/Encyclopedia/Meiklejohn.html
11. The Women's College in Brown University was established in 1891 but the women were separated from the men both socially and academically.

12. Nelson (2001). *Education and Democracy*, p. 57.
13. Nelson (2001). *Education and Democracy*, p. 39.
14. John Thelin, *Games Colleges Play*: Scandal and Reform in Intercollegiate Athletics (1986). Baltimore, MD: Johns Hopkins Press.
15. Mitchell (1993) Alexander Meiklejohn in *Encyclopedia Brunoniana*.
16. Wm. Alderman,(1928). "The Role of the Dean of Men in the Smaller Institution," *Secretarial Notes of the National Association of Deans of Men, 1928*, pp. 41–47.
17. Beloit College archives, William Alderman papers, File 1.
18. Ibid.
19. Alderman letter to Maurer, April 29. 1935, Beloit College Archives, Alderman papers, File 1.
20. Alderman obituary, April 19, 1977. Beloit College Archives, Alderman papers, File 1.
21. Alderman letter to Maurer, March 16, 1928, Beloit College Archives, Alderman papers, File 1.
22. Ibid.
23. Thomas Blayney (1928). "College Deans and the Report of the Carnegie Commission." *Secretarial Notes of the 10th Annual Meeting National Association of Deans of Men* pp. 21–30.
24. Blayney, Carleton College Archives, Northfield, MN.
25. Maurice Mendelbaum, Harold Speight Memorial Minutes, Proceedings of the American Philosophical Association, Volume 49 (1975–1976), pp. 163–164.
26. Thomas Lindsey Blayney records, Carleton College Archives, Northfield, MN.
27. Thomas Lindsey Blayney records, Carleton College Archives, Northfield, MN.
28. Maurice Mandelbaum, Memorial Minutes, Proceedings and Addresses of the American Philosophical Society, Volume 49, (1975–1976), p. 16.
29. University of California, Berkeley website. Retrieved September 21, 2009 from http://grad.berkeley.edu/lectures/foerster/pastlec.shtml
30. H. Hawkes and A. Hawkes (1945). *Through A Dean's Open Door*. New York, McGraw Hill. p. 8.
31. Although Hawkes was dead by the time the book was published, Anna Hawkes insisted that he had reviewed all but the last chapter and thought it ready for publication. This quote is an accurate representation of Hawkes' view of college life.
32. J. A. Fley. Student Personnel Pioneers: Those who developed our Profession. *NASPA Journal*, 17, (1979, June), pp. 41–44.
33. Fley, p. 42.
34. Fley, p. 42.
35. The Student Personnel Point of View, ACE, 1937.
36. Zook, Preface, Through A Dean's Open Door, 1937, New York: McGraw Hill.

37. Scott, www.library.northwestern.edu/archives/exhibits/presidents/scott.html
38. Ibid, Northwestern University archives, Scott bio.
39. Scott ref.
40. Scott, Northwestern University archives.
41. Scott becomes pres, of Northwestern, Ibid.
42. Hopkins and Lloyd-Jones join Scott Scott, www.library.northwestern.edu/archives/exhibits/presidents/scott.
43. National Association of Deans of Men minutes, 1926.
44. W. D. Scott papers, Northwestern University archives.
45. Cowley Memorial (obituary), Stanford University (1978), composed by L.G. Thomas and Lewis Mayhew. (retrieved on line 1/17/10 from histsoc.stanford.edu/pdfmem/**Cowley**W.pdf
46. Ibid., Cowley left Ohio State in 1938 to become president of Hamilton College (NY). In 1945, he joined the faculty at Stanford as the David Jacks Professor of Education. He retired in 1968 and died in 1978 in Palo Alto, CA.
47. Northwestern University archives, Presidents, Walter Dill Scott (online).
48. W. H. Cowley, "The Disappearing Deans of Men," *Secretarial Notes of the 19th Annual Meeting of the NADM, 1937*, pp. 85–99.

5 Francis F. Bradshaw: A Southern Student Personnel Pioneer

1. *Secretarial Notes of the National Association of Deans of Men, 1931.* Lawrence, KS: Republican Print.
2. Joseph Bursley, In *Secretarial Notes of the National Association of Deans of Men, 1931.* Lawrence, KS: Republican Print., p. 103.
3. Bursley, p. 104.
4. F. F. Bradshaw, *Secretarial Notes of the National Association of Deans of Men, 1931.* Lawrence, KS: Republican Print. p. 108.
5. ACPA Presidents (biography) Francis F. Bradshaw, Fourth President, (ACPA) 1928–1930, Bowling Green State University archives, www.bgsu.edu/colleges/library/cac/sahp/pdfs/bradshaw.pdf
6. *Secretarial Notes of the National Association of Deans of Men, 1928,* Lawrence, KS: Republican Print. p. 40.
7. E. Lloyd-Jones (1929). *The Student Personnel Movement at Northwestern University.* New York: Harper and Brothers Publishers.
8. Walter Dill Scott biography, Northwestern University Archives, The Presidents of Northwestern. www.library.northwestern.edu/archives/exhibits/presidents/scott.html
9. W. Urban & J. Wagoner (2008). *American Education: A History. New York: McGraw Hill.*
10. L. R. Wilson (1956). *The University of North Carolina, 1900–1930: The making of a modern university.* Chapel Hill, NC: The University of North Carolina Press.

11. Wilson, Ibid.
12. J. Edgerton (1994). *Speak Now Against the Day: The generation before the Civil Rights movement in the South. Chapel Hill, NC: University of North Carolina Press*, p. 130.
13. Ibid., p. 130, 131.
14. L. R. Wilson (1956). *The University of North Carolina, 1900–1930: The making of a modern university*. Chapel Hill, NC: The University of North Carolina Press. Wilson, (1956) The University of North Carolina.
15. Scott Goodnight, in *Secretarial Notes of the NADM*, 1934, p. 55.
16. F. F. Bradshaw (1931). In *Secretarial notes of the 13th annual conference of the National Association of Deans of Men, 1931*. Lawrence, KS: Republican Printing. pp. 108.
17. Francis F. Bradshaw, Fourth President, (ACPA) 1928–1930, Bowling Green State University archives, www.bgsu.edu/colleges/library/cac/sahp/pdfs/bradshaw.pdf
18. NAAS and ACPA, (need ref.)
19. Robert Rienow (1928) In *Secretarial notes of the 10th annual conference of the National Association of Deans of Men, 1928*. Lawrence, KS: Republican Printing, p. 28.
20. *Secy' Notes of the NADM, 1925*.
21. Appendix E, Summary of previous meetings. *Secretarial notes of the 16th Annual Conference of the National Association of Deans of Men, 1934*. Lexington, KY: University of Kentucky Press. p. 158.
22. L. R. Wilson (1956) *The University of North Carolina, 1900–1930: The making of a modern university*. Chapel Hill, NC: The University of North Carolina Press.
23. L. R. Wilson, ibid
24. R. Schwartz (1997). "How Deans of Women Became Men." *Review of Higher Education, 20*.
25. L. R. Wilson, Ibid.
26. Ibid.
27. U.S. Commissioner of Education, (1919) *Report of the U.S. Commissioner of Education, 1916–1918*, U.S. Government Documents, p. 65.
28. *Report of the U.S. Commissioner of Education, 1931*, pp. 420–424.
29. L. R. Wilson, ibid
30. ACPA President's (biography) Francis F. Bradshaw, Fourth President, (ACPA) 1928–1930, Bowling Green State University archives, www.bgsu.edu/colleges/library/cac/sahp/pdfs/bradshaw.pdf
31. Bradshaw, F.F. (1939) In *Secretarial notes of the 21st annual conference of the National Association of Deans of Men, 1939*. Lawrence, KS: Republican Printing. pp. 30–31.
32. Mark L. Savickas and David L. Baker (2005). A History of Vocational Psychology: Antecedents, Origin, and Early Development. In Handbook of Vocational Psychology, Third Edition, Mahwah, NJ, Lawrence Erlbaum Publishing, pp. 15–50.
33. Ibid.

34. Ibid.
35. R. A. Schwartz (1997) Reconceptualizing the Leadership Roles of Women in Higher Education: A Brief History on the Importance of Deans of Women (Fall, 1997) *Journal of Higher Education.*
36. Ibid.
37. Max McConn (1931). The Co-Operative Test Service: A history of the Co-operative Test Service of American Council of Education and the Proposed Nation-wide Testing of College Sophomores. *The Journal of Higher Education*, pp. 225–232.
38. Ibid., p. 226.
39. Ibid., p. 226.
40. L. B. Hopkins, (1926) Personnel procedures in higher education: Observations and conclusions resulting from visits to fourteen institutions of higher learning. New York: American Council on Education.
41. W.H. Cowley (1937) The disappearing dean of men. In Secretarial notes of the 17th annual meeting of the National Association of Deans and Advisers to Men. Lawrence, KS: Republican Printing, p. 85–99.
42. Bradshaw was not the only member on the committees from the University of North Carolina. M. R. Trabue, a professor of educational administration, was listed as a member of the sub-committee on achievement tests. His record in the UNC archives shows Trabue on leave from the university in the Fall and Winter terms of 1927, 1931/32, and 1932/33, and again in Winter and Spring of 1936. See Register of the Officers and Faculty of the University of North Carolina, 1795–1945 at: www.lib.unc.edu/ncc/ref/unc/faculty.html
43. A Nursery of Patriotism, UNC Archives. www.lib.unc.edu/mss/exhibits/patriotism/index.php?page=WorldWarII-Preparing&size=small
44. ACPA President's (biography) Francis F. Bradshaw, Fourth President, (ACPA) 1928–1930, Bowling Green State University archives, www.bgsu.edu/colleges/library/cac/sahp/pdfs/bradshaw.pdf
45. American Council on Education (1937) *The Student Personnel Point of View.* Author: Washington, D.C.
46. *Secretarial Notes of the National Association of Deans of Men, 1931,* Raleigh, NC: Edwards & Broughton Company, p. 27.

6 A Modern Dean: Fred Turner

1. "An outline of the staff members of the Office of Dean of Men at the University of Illinois from 1901 to 1937. T.A. Clark papers. UISLA. Box 41/2/20.
2. Oral history interview with Turner by Jennifer Johnson, WILL (radio), July 25, 1967. UISLA. Box 41/1/20 UISLA. Box 41/1/20.
3. Ibid.
4. Fred Turner papers, UISLA. Box 41/1/20.

5. Ibid.
6. Oral history interview with Turner by Jennifer Johnson, WILL (radio), July 25, 1967. UISLA. Box 41/1/20 UISLA. Box 41/1/20.
7. Turner papers, UISLA. Box 41/1/20.
8. An outline of the staff members of the Office of the Dean of Men at the University of Illinois from 1901 to 1937. Clark papers, UISLA. Box 41/2/20.
9. Ibid.
10. T.A. Clark, How to Obtain Personal Contact in Large Universities. In Secretarial Notes on the Eighth Annual Conference of Deans and Advisers of Men. University of Minnesota, 1926. Raleigh, NC: Edwards and Broughton Company. pp. 46–48.
11. Appendix A, Secretarial Notes on the Eighth Conference of Deans and Advisers of Men. University of Minnesota, 1926. Raleigh, NC: Presses of Edwards and Broughton Company. pp. 61.
12. Ibid.
13. T.A. Clark, How to Obtain Personal Contact in Large Universities Secretarial Notes of the Eighth Annual Conference of Deans and Advisers of Men, 1926. Raleigh, NC: Presses of Edwards & Broughton, pp. 46–48.
14. Fred Turner, Report of the Committee on Preparation for Work As A Dean of Men. *In Secretarial Proceedings of the 18th Annual Conference of the NADM,* Philadelphia, PA, 1936, Lawrence, KS: Republican Press, pp. 16–52.
15. Ibid.
16. Fred Turner, Report of the Committee on Preparation for Work As A Dean of Men. In *Secretarial Proceedings of the 18th Annual Conference of the NADM*, Philadelphia, PA 1936, Lawrence, KS: Republican Press, pp. 16–52.
17. T. A. Clark, The Relation of the College Faculty to Fraternities. *Journal of the Proceedings and Addressses of the Forty-eighth Annual Meeting. National Education Association of the United States, 1910.* Winona, MN: NEA. pp. 548–557.
18. R. Clothier, cited by F. Turner, Report of the Committee on Preparation for Work As A Dean of Men. In *Secretarial Proceedings of the 18th Annual Conference of the NADM*, Philadelphia, PA, 1936, Lawrence, KS: Republican Press, pp. 17–18.
19. Fred Turner papers, UISLA. Box 41/1/20.
20. Oral history interview with Turner by Jennifer Johnson, WILL (radio), July 25, 1967. UISLA. Box 41/1/20 UISLA. Box 41/1/20.
21. Ibid.
22. Fred Turner, Office Staff, Records, and Organization In the Dean of Men's Office. Secretarial Proceedings of the 12th Annual conference of NADAM, Fayetteville, TN, 1930. Lawrence, KS: Republican Press, pp. 80–84.

23. R. Clothier, The Relation of the Dean of Men to Personnel Work in the Larger University. In *Secretarial Notes of the 13th Annual Conference of Deans and Advisers of Men,* Gatlinburg, TN, 1931. Lawrence, KS: Republican Press, pp. 29–38.
24. Ibid.
25. F. F. Bradshaw cited by Turner, Report of the Committee on Preparation for Work As A Dean of Men. In *Secretarial Proceedings of the 18th Annual Conference of the NADM*, Philadelphia, PA 1936, Lawrence, KS: Republican Press, p. 22.
26. J. Armstrong, cited by F. Turner, Report of the Committee on Preparation for Work As A Dean of Men. In *Secretarial Proceedings of the 18th Annual Conference of the NADM*, Philadelphia, PA, 1936, Lawrence, KS: Republican Press, pp. 22–23.
27. D. Gardner cited by Turner, Report of the Committee on Preparation for Work As A Dean of Men. In *Secretarial Proceedings of the 18th Annual Conference of the NADM*, Philadelphia, PA 1936, Lawrence, KS: Republican Press, pp. 24–25.
28. Fred Turner, Report of the Committee on Preparation for Work As A Dean of Men. In *Secretarial Proceedings of the 18th Annual Conference of the NADM*, Philadelphia, PA 1936, Lawrence, KS: Republican Press, pp. 16–52.
29. Turner, Ibid.
30. Ibid.
31. Ibid.
32. Ibid.
33. Ibid.
34. Ibid.
35. H. Speight, *Secretarial Notes of the 18th Annual Conference of Deans and Advisers of Men*, Phildelphia, PA, 1936. Lawrence, KS: Republican Press, p. 49.
36. F. M. Massey, *Secretarial Notes of the 18th Annual Conference of Deans and Advisers of Men*, Philadelphia, PA 1936. Lawrence, KS: Republican Press, pp. 50–51.
37. Unidentified speaker, discussion related to Turner's report. *Secretarial Notes of the 18th Annual Conference of Deans and Advisers of Men*, Philadelphia, PA 1936. Lawrence, KS: Republican Press, p. 50–51.
38. Digest of Educational Statistics. National Center for Educational Statistics (1971–1985, inclusive) Washington, D. C, : U.S. Government Printing Office.
39. L. B. Hopkins, The Nature and Scope of Personnel Work. In *Secretarial Notes of the 13th Annual Conference of Deans and Advisers of Men*, Gatlinburg, TN, 1931. Lawrence, KS: Republican Press, pp. 22–29.
40. R. Clothier, The Relation of the Dean of Men to Personnel Work in the Larger University. In *Secretarial Notes of the 13th Annual Conference of Deans and Advisers of Men,* Gatlinburg, TN, 1931. Lawrence, KS: Republican Press, pp. 29–38.
41. Ibid, p. 32.

42. Ibid, p. 36.
43. Ibid, p. 36.
44. Lorin Thompson, Jr. The Relation of the Personnel Officer to the Dean of Men in the Small College. In *Secretarial Notes of the 13thAnnual Conference of Deans and Advisers of Men*, Gatlinburg, TN, 1931. Lawrence, KS: Republican Press, pp. 38–44.
45. A Survey of Land–Grant College and Universities, in Report of the U.S. Commissioner of Education, 1932. Discipline of Women and Male Students, 1928. in A Survey of Land-Grant Colleges and Universities. In Report of the U.S. Commisionher of Education, 1932. Washington, D.C.: U.S. Government Printing Office.
46. W. H. Cowley, The Disappearing Deans of Men. *Secretarial Notes of the 19th Annual Conference of Deans and Advisers of Men*, Austin, TX, 1937. Lawrence, KS: Republican Press, pp. 85–99.
47. Ibid., p. 93
48. Ibid., p. 94.
49. Ibid., p. 94.
50. Ibid., p. 97.
51. Ibid., p. 98.
52. Ibid., p. 99.
53. Ibid., p. 99.
54. *The Student Personnel Point of View* (1937). New York: American Council on Education retrieved from www.myacpa.org/pub/documents/1937.pdf
55. J. F. Findlay, "Origins and Development of the Work of the Dean of Men," *Secretarial Notes of the 19th Annual Conference of Deans and Advisers of Men*, Austin, TX, 1937. Lawrence, KS: Republican Press, pp. 104–121.
56. Ibid., pp. 107–108.
57. Ibid., pp. 109–110.
58. Ibid., p. 112.
59. Walter Dill Scott, cited by Findlay, p. 119
60. F. Turner, *Secretarial Notes of the 19thAnnual Conference of Deans and Advisers of Men*, Austin, TX, 1937. Lawrence, KS: Republican Press, pp. 121–122.
61. Clark, cited by Fred Turner, *Secretarial Notes of the 19thAnnual Conference of Deans and Advisers of Men*, Austin, TX, 1937, pp. 122.
62. D.H. Gardner, Survey of the Functions of the Dean of Men, *Secretarial Notes of the 14th Annual Conference of Deans and Advisers of Men*, Los Angles, CA, 1932. Lawrence, KS: Republican Press.
63. F. F. Bradshaw, *Secretarial Notes of the 21st Annual Conference of Deans and Advisers of Men*, 1939. Lawrence, KS: Republican Press, pp. 30–31.
64. D.H. Gardner, *Secretarial Notes of the 21st Annual Conference of Deans and Advisers of Men*, 1939. Lawrence, KS: Republican Press, p. 46. (not sure what is needed here? It says "need footnote" but the note is there?
65. Fred Turner papers, University of Illinois Student Life Archives, 41/1/20.

7 A Brief Treatise on the Deans of Women

1. Irma Voight (1936). *Yearbook of the National Association of Deans of Women.*
2. Jana Nidiffer (2000). *Pioneering Deans of Women: More than Wise and Pious Matrons.* New York: Teachers College Press; Carolyn Bashaw (1999) *Stalwart Women: A Historical Analysis of Deans of Women in the South.* New York: Teachers College Press.
3. Kathryn Nemeth Tuttle (1996). What Became of the Deans of Women? Changing Roles for Women Administrators in American Higher Education. Unpublished doctoral dissertation, University of Kansas; Lynn Gangone (1999). Navigating Turbulence: A Case Study of a Voluntary Higher Education Association. Unpublished doctoral dissertation, Teachers College, Columbia University; and R. A. Schwartz (1990). The Feminization of a Profession: Student Affairs Work in American Higher Education, 1890–1945. Unpublished doctoral dissertation, Indiana University.
4. R. Schwartz (1997). How Deans of Women Became Men. *The Review of Higher Education* 20(4) (Summer), pp. 419–436.
5. M. Talbot (1925). *More Than Lore: Reminiscences of Marion Talbot.* Chicago: University of Chicago Press.
6. Schwartz, How Deans of Women Became Men.
7. Barbara Solomon (1985). *In the Company of Educated Women: A History of Women and Higher Education in America.* New Haven, CT: Yale University Press, 7.
8. Ibid.
9. Talbot, *More Than Lore.*
10. Robert Schwartz (1997). Reconceptualizing the Leadership Roles of Women in Higher Education: A Brief History on the Importance of Deans of Women. *The Journal of Higher Education*, 68(5) (Sep.–Oct.), pp. 502–522.
11. National Association of Deans of Women, *Yearbook of the National Association of the Deans of Women*, 1927.
12. Schwartz, How Deans of Women Became Men.
13. Ibid.
14. Lois K. Mathews (1915). *The Dean of Women.* Cambridge, MA: The Riverside Press.
15. Marion Talbot (1910). *The Education of Women.* Chicago: The University of Chicago Press.
16. M. Talbot and L. (Mathews) Rosenberry (1936). *The History of the American Association of University Women.* Cambridge, MA: The Riverside Press.
17. Anna Pierce (1929). *Deans and Advisers of Women and Girls.* New York: Professional and Technical Press.
18. Sarah Sturtevant and Ruth Strang (1928). *A Personnel Study of Deans of Women in Teachers Colleges and Normal Schools.* Contributions

to Education, No. 319. New York: Teachers College, Columbia University.

19. Sarah Sturtevant and Ruth Strang (1929). *A Personnel Study of Deans of Girls in High Schools*. Contributions to Education, No. 393. New York: Teachers College, Columbia University.
20. Schwartz, Reconceptualizing the Leadership Roles of Women in Higher Education.
21. National Association of Deans of Women. *Yearbooks of the National Association of the Deans of Women.1923–1940.*
22. R. Strang (1934) Report of the Research Committee. *Yearbook of the National Association of Deans of Women*, pp.56
23. C. Burgess (1971). In Havighurst, R.J. (Ed.) Leaders in American Education. *The 17th Yearbook of the National Society for the Study of Education, Part II*. Chicago: University of Chicago Press, pp. 938–411.
24. R. Strang. In National Association of Deans of Women. *Yearbook of the National Association of the Deans of Women*, .
25. Eunice M. Acheson (1932). The *Effective Dean of Women*. New York: Teachers College Press.
26. Ibid.
27. Ibid.
28. Ibid
29. P. Graham (1978). Expansion and Exclusion: A History of Women in American Higher Education, *Signs: Journal of Women in Culture and Society*. (3), pp. 759–773.
30. Sarah Sturtevant, Ruth Strang, and McKim (1940). *Trends in Student Personnel Work,* Contributions to Education, #787. New York: Teachers College, Columbia University.
31. Nancy Cott (1987). *The Grounding of Modern Feminism*. New Haven, CT: Yale University Press, p. 217. See also Solomon, *In the Company of Educated Women*; William Chafe (1977). *The American Woman: Her Changing Social, Economic, and Political Roles, 1920–1970*. New York: Oxford University Press; Frank Strickler (1984). Cookbooks and Law Books. In Katz, A. & Rapone, A. (Eds.) *Women's Experience in America: An Anthology*. New Brunswick, NJ: Transaction Press; and Lois Scharf (1980) *To Work and To Wed*. Westport, CT: Greenwood Press.
32. Lois Scharf and Joan Jensen (Eds.) (1983). *Decades of Discontent: The Women's Movement, 1920 to 1940 (Contributions to Womens Studies)*. Westport, CT: Greenwood Press. p. 13.
33. Ibid., p. 92.
34. Ibid., p. 190.
35. Lois Scharf, *To Work and To Wed*.
36. Cott, p. 217
37. Ibid.
38. Ibid.

39. U. S. Office of Education, Department of the Interior (1931) Survey of Land Grant Colleges and Universities, Bulletin No. 9, (Volumes (I and II). Washington, D.C.: Superintendant of Documents.
40. R. Schwartz (1997) How Deans of Women Became Men.
41. Ibid.
42. W. H. Cowley (1937). *Secretarial Notes of the 19th Annual Conference of the National Association of Deans and Advisers of Men NADAM 1937*, Austin, Texas, pp. 85–101.
43. Schwartz, How Deans of Women Became Men.
44. Ibid
45. Cowley, *Secretarial Notes.*
46. John Thelin (2004). *The History of American Higher Education* Baltimore: Johns Hopkins Press.
47. Horowitz, (1987) *Campus Life: Undergraduate Cultures from the Eighteenth Century to the Present.* Chicago: University of Chicago Press.
48. R. Schwartz (1998). Lessons from the Past: Women in Higher Education leadership in the Depression years. *Initiatives: The Journal of the National Association for Women in Education, 58*, 1–21
49. Thelin. *The History of American Higher Education.*
50. Sarah Sturtevant, Ruth Strang, and McKim (1940). *Trends in Student Personnel Work,* Contributions to Education, #787. New York: Teachers College, Columbia University.
51. J. F. Findlay (1937) The Origins and Development of the Work of the Dean of Men. Secretarial Notes of the 19th Annual Conference of Deans and Advisers to Men (NADAM) Austin, TX, pp. 104–121
52. Sarah Sturtevant, Ruth Strang, and McKim (1940). *Trends in Student Personnel Work,* Contributions to Education, #787. New York: Teachers College, Columbia University.
53. Ibid.
54. Ibid.
55. Ibid.
56. Ibid.
57. Ibid.
58. Ibid.
59. Ibid.
60. Ibid.
61. Ibid.
62. Ibid.
63. Ibid.
64. Ibid.
65. Ibid.
66. Ibid.
67. Ibid.
68. Schwartz, Lessons from the Past.
69. Sturtevant, Strang, and McKim, *Trends in Student Personnel Work,* p. 52
70. Ibid, p. 55

71. Ibid.
72. Esther Lloyd-Jones (1940) *Social Competence.* Washington, DC: American Council on Education. p. ix.
73. Ibid.
74. Ibid., p. x.
75. Sturtevant, Strang, and McKim, *Trends in Student Personnel Work*, p. 95
76. Scharf and Jensen (Eds.), *Decades of Discontent*
77. Ibid., p. 13.
78. Louise Spencer (1952), Eleven Years of Change in the Role of Deans of Women in Colleges, Universities, and Teachers Colleges. Unpublished doctoral dissertation, Teachers College, Columbia University, New York.
79. Solomon, *In the Company of Educated Women* p. 185.
80. Ibid., p. 185.
81. Scharf and Jensen (Eds.) (1983). *Decades of Discontent.*
82. Ibid., p. 13.
83. Spencer (1952). Eleven Years of Change in the Role of Deans of Women in Colleges, Universities, and Teachers Colleges. Unpublished doctoral dissertation, Teachers College, Columbia University, New York.
84. Spencer, Eleven Years of Change, p. 216.
85. Ibid., pp. 269–270
86. Ibid., pp. 116–117.
87. Ibid., p. 112.
88. Ibid., pp. 112–113.
89. Michael Coomes, Elizabeth Whitt, and George Kuh (1987). Kate Hevner Mueller: Woman for a Changing World. *Journal of Counseling and Development*, 65 (8), pp. 407–415.
90. August Eberle cited by Coomes, et al.
91. William Chafe (1972). *The American Woman: Her Changing Social, Economic, and Political Roles, 1920–1970.* New York: Oxford University Press. p. 180.
92. Ibid., p. 308.
93. Ibid., pp. 205–206.
94. Graham. Expansion and Exclusion.
95. Solomon, *In the Company of Educated Women*, pp. 191–192.
96. American Council on Education (1937), "The Student Personnel Point of View."
97. Solomon, *In the Company of Educated Women*, pp. 191–192.
98. American Council on Education, "The Student Personnel Point of View."
99. Schwartz, How Deans of Women Became Men
100. Ibid.
101. Chafe, *The American Woman.*
102. Louise Spencer (1952), Eleven Years of Change in the Role of Deans of Women in Colleges, Universities, and Teachers Colleges. Unpublished doctoral dissertation, Teachers College, Columbia University, New York.
103. Graham Expansion and Exclusion.

104. Solomon, *In the Company of Educated Women*, pp. 191–192
105. Schwartz, How Deans of Women Became Men.
106. Chafe, *The American Woman*.

8 The Demise of the Deans of Men and the Rise of the Deans of Students

1. Proceedings of the Thirty-first Anniversary Conference of the National Association of Deans and Advisers to Men, 1949, p. 192.
2. Proceedings of the Thirty-first Anniversary Conference of the National Association of Deans and Advisers to Men, 1939.
3. Proceedings of the Thirty-first Anniversary Conference of the National Association of Deans and Advisers to Men, 1949, pp. 78
4. J. H. Newman, in Proceedings of the Thirty-first Anniversary Conference of the National Association of Deans and Advisers to Men, 1949, pp. 8–17.
5. Ibid.
6. Ibid.
7. Ibid., p. 9.
8. Ibid., p. 10.
9. Ibid., p. 11.
10. Ibid., p. 13.
11. ibid., p. 14.
12. Proceedings of the Thirty-first Anniversary Conference of the National Association of Deans and Advisers to Men, 1949, pp. 139–150. There is considerable discussion of the same issues related to the Association articles and the name change in the Proceedings of the 30th Anniversary Conference held in Dallas, TX, on pages 173–180.
13. Proceedings of the Thirtieth Anniversary Conference of the National Association of Deans and Advisers to Men, 1948. Dallas, TX.
14. Donald M. DuShane in Proceedings of the Thirty-first Anniversary Conference of the National Association of Deans and Advisers to Men, 1949, p. 144.
15. Ibid., p. 144.
16. Arno Nowotny in *Proceedings of the Thirty-first Anniversary Conference of the National Association of Deans and Advisers to Men*, 1949, p. 150.
17. Wesley Lloyd, in *Proceedings of the Thirty-third Anniversary Conference of the National Association of Student Personnel Administrators*, 1951, p. 12.
18. Ibid., p. 13.
19. Ibid., p. 13.
20. Ibid., p. 15.
21. D. Newhouse in *Proceedings of the Thirty-third Anniversary Conference of the National Association of Student Personnel Administrators*, 1951, p. 19.

22. Blair Knapp in *Proceedings of the Thirty-third Anniversary Conference of the National Association of Student Personnel Administrators*, 1951, p. 22
23. Ibid., p. 23.
24. Appendix A, Report of the Secretary, *Proceedings of the Thirty-third Anniversary Conference of the National Association of Student Personnel Administrators*, 1951, p. 206.
25. A. French, in *Proceedings of the Thirty-third Anniversary Conference of the National Association of Student Personnel Administrators*, 1951, p. 37.
26. E. G. Williamson in *Proceedings of the Thirty-third Anniversary Conference of the National Association of Student Personnel Administrators*, 1951, p. 62.
27. Ibid.
28. Wayne Spathelf in *Proceedings of the Thirty-third Anniversary Conference of the National Association of Student Personnel Administrators*, 1951, p. 150.
29. *Proceedings of the Thirty-third Anniversary Conference of the National Association of Student Personnel Administrators*, 1951, p. 152.
30. Victor Spathelf, ibid., p. 153.
31. Wesley Lloyd, ibid., p. 154.

9 A Retrospective Epilogue

1. Irma Voight (1936) Yearbook of the NADW, 1936
2. Brown, (1936) *Dean Briggs*. New York: Harper and Sons.
3. Voight, Yearbook of the NADW.
4. Laurence Veysey (1965), Emergence of the American University. Chicago: University of Chicago Press.
5. Stanley Coulter (1928). Secretarial Notes of the National Association of Deans and Advisers to Men.
6. Clark obituary, *New York Times* (1931).
7. Scott Goodnight (1934).
8. For more on early discipline issues, see Kathryn Moore's (1976) Freedom and Constraint in Eighteenth Century Harvard, *The Journal of Higher Education*, 47(6), pp. 649–659, and Jennings Wagoner (1986). Honor and Dishonor at. Mr. Jefferson's University *History of Education Quarterly*, 26(2), pp. 155–179.
9. Secretary's Report, Proceedings of the National Association of Student Personnel Administrators, 1952.
10. Walter Dill Scott biography, Northwestern University Archives.
11. Roger Geiger (1993). *Dissolution of Consensus, Research and Relevant Knowledge: American Research Universities since World War II*. New York: Oxford University Press.

12. M. Lee Upcraft and John N. Gardner (1989). *The Freshman Year Experience: Helping Students Survive and Succeed in College.* San Francisco, CA: Jossey-Bass Publishers.
13. Alexander Astin (1993). *What Matters in College: Four Critical Years Revisited.* San Francisco, CA: Jossey-Bass Publishers.
14. George Kuh (2001). National Survey of Student Engagement- The College Student Report. NSSE Technical and Norms Report. Bloomington, IN: Center for Postsecondary Research and Planning.

Index